PARASITE ZAPPING AND THE ZAPPER

Fighting Disease and Illness with mild,
safe Electric pulses.

DAVID L. ETHEREDGE

IMPORTANT: Read this manual completely and thoroughly!

> **This book is dedicated to my former wife Alice, for her devotion to my efforts to help others. She succumbed to lung cancer on 13 June 2006.**

Acknowledgements

I would like to thank those who have contributed including my current wife Lydia for her help with this, former wife Alice Etheredge, my mother Frances Etheredge, and my sister Dr. Melinda Carter M.D.

I also extend a special thank you to Dr. Hulda Regehr Clark without whom the zapper would probably not exist.

Introduction

The information contained in this booklet is presented for those who are seeking to investigate the mysteries of electricity in relation to water and to bodies made mostly of water. By necessity, the most important body of water to the human reader is the human body itself. This publication is not intended to be for use with any particular product but is instead intended to open the doors of thought in regards to a range of products.

I first heard of using the zapper to cure diseases from Wayne Green of magazine publishing fame. In an editorial, he expanded on the virtues of the zapper and its ability to cure diseases. The zapper that he was discussing was the Bob Beck zapper, otherwise known as the Bob Beck Blood Electrifier (BBBE).

```
Admittedly,  I  was  very  skeptical  of  this  because,
after  all,  I  had  years  of  Biology,  Chemistry,  and
electronics  and  none  of  this  had  ever  even  been
mentioned.  I  had  even  had  graduate  school  in
biomedical  engineering  and  it  was  never  mentioned.
Still, I found the idea intriguing and set the article
aside for later investigation.
```

For a number of years, I heard nothing of the zapper until one winter and spring, a number of the people in the office where I worked started coming down with the flu. Everyone was missing 3 to 5 days and still not doing well. This continued for a few weeks, with the cycle repeating itself. The same people, including myself were getting sick over and over again. One person in the office appeared not to be catching the flu. Then one afternoon, he suddenly came down with the bug, leaving to go home. He returned the next day, feeling fine!

When I asked him his secret, he handed me a little hand built zapper, claiming that it helped him stay healthy. Immediately impressed, I decided to try one and was amazed with the results. I borrowed one of my friends zappers and took it home. A few nights later, I suddenly felt the flu coming on again. I picked up the zapper and used it for about an hour, after which, I felt tired and felt like it did not do anything for me. I went to bed and the next

1

morning when I got up, I felt the best that I had felt in years. It definitely increased my energy levels and there was no symptom of the flu.

Later, using the zapper, I removed the pain and swelling of a tooth infection in a day, broke colds and the flu, stopped a bad fungal infection, reduced Candida, and have felt healthier since starting to use it. I also noticed later that my problem with IBS had been helped. As time progressed, I started making more advance zappers and did a number of experiments. In late 2003, I experimented with the ParaZapper MX and found it to be much better. In 2004, I used it along with a flush, passing a tapeworm that I had for most of my life. Using it, I also got rid of athletes foot infections that medications did not help. Later that year, I found a tick on my abdomen and the bite soon started developing a large bulls-eye rash that is typical of Lyme disease. I zapped this for several days in a row and it went away. I have not had any symptoms since. Since then, I have stopped several abscessed teeth, stopped a serious deep wound infection, and used it for a number of other situations, including spider bites. I also used it to destroy several secondary lesions of basal cell cancer and have remained cancer free for over 10 years now. Also, As a child and in my adult years up to the time that I started using the zapper, I had frequent oral herpes but this went away after I started using the zapper and I have never had another outbreak.

Use of the zapper was first suggested by Dr. Hulda Clark, in her book "The Cure for All Diseases". I recommend that everyone read this along with her other books.

Not being a doctor, I cannot prescribe the zapper to you, nor can I guarantee any results, but I do believe in its effectiveness in reducing the effects of illness. From the examples above, I have had tremendous results from using these amazing devices but that does not prove anything other than my own satisfaction and results..

We do frequently have customers call us back to tell us about how ParaZapper has improved their health and about the parasites that they have passed.

While the ParaZapper may help with many parasitic infections, do not fool yourself into thinking that zapping once or twice or even just a few times will do the job. **The more you zap the better your results will be in general.** Killing parasites takes work and time. It is also important to understand that the zapper is not a cure all but it is one tool that can help you recover from failing health or help you to avoid the future problems associated with failing health. **See:** Intestinal Flora, Foods and Nutrition, Zapping and Metals.

It is also important to read about how to zap for the best results as well as things not to do when zapping. Studying this information can make your zapping experience more pleasurable; produce better results, and may save you some misery.

The zapper is not entirely new and did not begin with Dr. Clark. Devices that operate on the same basis have been around for over 150 years. There are relevant patents dating back to 1859 and such devices were found on the shelves in drug stores and even in the Montgomery Ward catalog and the Sears, Roebuck catalog. The history of using electricity as a cure for illness goes back to the middle of the 1700's and the medical profession has been trying to keep them from usage ever since.

These devices were forced off the market in the 1930's when the AMA was trying to consolidate its power. There have been numerous devices developed along the way such as the Rife machine and the Beck Blood electrifier. A significant amount of work and research was performed from the 1950's through the 1970's by Dr. Robert O. Becker, who showed that electricity as low as 1.5 volts could not only kill bacteria living in the body, but that it could also promote and speed healing, including in cases of bone fractures. His story was published in "Body Electric" and "Cross Currents" and both of these are excellent reading for interested parties.

The idea that an electrical voltage can help heal and can take the place of antibiotics and anti-virals in some cases, is a major threat to the medical profession. Rather than support these technologies

3

and the humanitarian gains made through them, the doctors and their public representative, the AMA, consistently try to oppress this information and to deny the public access.

The Russian astronauts have for over 50 years taken an electric device, referred to as a SCENAR, with them into space to help keep them healthy and strong. In the 1920's and early 1930's. Royal Raymond Rife developed several scientific advancements, including several powerful optical microscopes that were as much as 20 times as powerful as what is generally used to day. He also developed a radio frequency device that cured cancer in mice and in a study, was effective in stopping cancer in humans. He attributed these cancers to a virus. Later, his labs were shut down and his machines and microscopes were mostly destroyed. Even today, the FDA is seeking to destroy the remaining examples of his technology.

The one thing that I can express through my few years of experience with zappers is that they do work in many cases and that over 90 percent of the users who buy one, recommend to their friends to get one. We have had customers whom after using the zapper for a few months or years, come back and buy more for their family and friends as gifts.

Zapper Kills Microbes

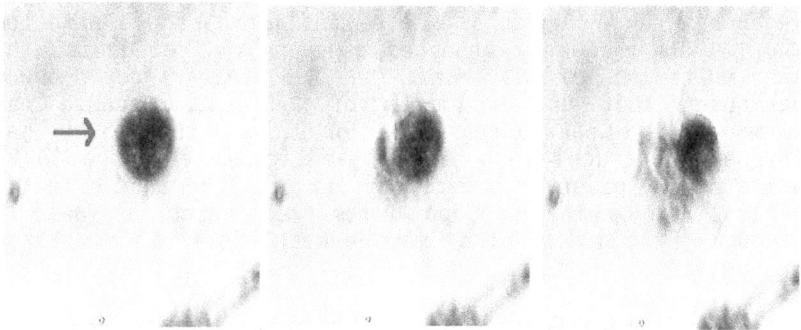

Actual protozoa breaking apart while being zapper.

The truth is that zappers do work but they are not perfect. We have been working hard to make improvements to make them more effective in more applications.

How does the zapper work?

The study of Radionics suggests that all organisms are susceptible to certain frequencies of electromagnetic energy. A vibration of 5 Hz will kill chickens due to resonance of their skull cavity. Many bacteria and viri are affected by frequencies of 300 kHz to 450 kHz. Protozoa, roundworms and flatworms are affected by frequencies between 350 kHz and 500 kHz. Molds appear to be affected by lower frequencies of around 80 kHz to 220 kHz. Individual species respond to specific frequencies within these ranges.
If you have any doubt about this idea of resonance contact the imaging department at your local hospital. MRI and some other imaging technologies are based on this same principal. For more, **See Frequency Notes**.

While there are many versions of zappers available, I have found that there can be significant differences among them. The most important difference is that your zapper should have multiple frequencies in different modes. Why multiple frequencies? Each frequency has different frequency spectra, which may affect various microbes differently. Also, cheaper zappers may produce weaker spectra and as a result, may not work as well.

Most zappers produce a square wave output. It is important that the square wave has a positive offset to it. This square wave is composed of all odd harmonics of the main frequency. These harmonics effectively stretch up into the megahertz range, which is where many microbes are affected. The 30 kHz frequency produces a range of frequencies that are multiples of 30 kHz. The 2.5 kHz range produces harmonics that are multiples of 2.5 kHz. Because each harmonic is weaker than the previous, the lower base frequency produces much smaller amplitudes than the 30 kHz units at any given frequency but can still be more effective because they will produce more frequencies as well as producing frequencies that may be much closer to the target frequencies.

According to information available from the study of Radionics, microbes such as anthrax, Chlamydia, Shigella, and E. coli have spectral responses between 392 and 398 kHz. This would be near the 13th harmonic of 30 kHz, and near the 157th and 159th harmonic of 2.5 kHz. The 13th harmonic has an amplitude value of 1/13th of the main frequency and the 157th harmonic has an amplitude value of $1/157^{th}$ of the main frequency. The trade-off of amplitude vs. closeness to the needed frequency has to be weighed in considering desired results.

It should be obvious that the closer a frequency is to the required harmonic, the better the expected results. Also, for any given target frequency, the stronger the harmonic, the better the expected results. For this reason, more frequencies are likely to produce better results.

The end result is that the zapper causes microbes to vibrate internally and this vibration eventually breaks them apart essentially killing them. This effect has been recorded numerous times under the microscope.

Zapper Theory

Some users believed that the higher the frequency, the greater the skin effect. Basically, higher frequencies tend to travel along the surfaces of the body and lower frequencies tend to penetrate better. For this reason, many believe that the lower frequencies are more effective against intestinal parasites, urinary infections, and other infections that tend to occur in the various body cavities.

It might just be that the 30 kHz frequency does not produce harmonics close enough to the parasite's frequency since its harmonics are 30 kHz apart while the lower frequency has harmonics that are only 2.5 kHz apart. Regardless of the reason, many users feel that the 2.5 kHz frequency is more effective.

There are others who feel that 2200 Hz is better than 2500 Hz, but I have not seen any evidence to support this. What I have seen is that certain specific frequencies are more effective against certain specific organisms and also that certain specific frequencies affect more organisms that other frequencies do. In the 1950's John Crane, working with Royal Rife showed that one particular frequency may be better for a specific illness or a particular type of problem.

In reality, using two frequencies is probably much better than using just one because you have twice the chance of hitting the right frequency or close to it. The use of 3, 4, or even more frequencies is likely to improve results. In the most expensive products such as Rife machines, you can select individual specific frequencies with great accuracy. Of course, this is if you can afford several thousands of dollars.

Otherwise, a really good multi-mode, multi-frequency zapper can be as accurate as Rife machines and possibly even produce better results.

It is actually possible to use only a 9-volt battery and tap it with your fingers. It is difficult and the results may not be quite as good, but it does produce results. Around the world, people in many different places will resort to various means to obtain electrotherapy. In many wilderness areas, safari guides will resort to using automobile batteries to help break down the venom from snake bites, scorpion stings, and the bites of venomous spiders.

In Indonesia, many people with parasites will intentionally lay across the rails of electric train tracks to obtain relief.

If you want to look at nature, the best example of using vibrational energy for healing is the cat and this may well be the source of their reputation of having nine lives. It should be noted that there are cases in which a cat having fallen over 50 stories, breaking every bone in their body, and having recovered. What could be more amazing?

What frequency should I use?

It was originally recommended that under most situations you should use 30 kHz. If this does not have the desired effect then try the 2500 Hz range.

When using a dual frequency model, some users will zap a full cycle at 30 kHz and follow with another full cycle at 2500 Hz. Others **zap 7 minutes at 30 kHz followed by 7 minutes at 2.5 kHz followed by a rest of 20 minutes.** This method appears to work best in older dual frequency zappers.

It is strongly recommended not to alternate frequencies with successive zappings. In other words, do not zap one zapping at one frequency and then use a different frequency for the next zapping in the same session. The reason for this is that each frequency that you use must be repeated 3 times minimum and should be used for 7 minutes minimum each zapping (actually, about 7-1/2 minutes is better). If you want to use two frequencies, zap 7 minutes (or more) at one frequency, followed by 7 minutes at the second frequency, and then a 20-30 minute break. This should be repeated 3 times minimum and can be repeated 4 or 5 times for even better results.

When a zapper specifies a frequency, you should consider that it may not be exact. Most zappers put out a frequency that is in the range of the specified frequency. Often, the frequency will be either +/- 5 % or +/- 10 percent depending on the components used. So a zapper specifying 30 kHz could have a frequency of 27 kHz or 33 kHz or anything in between. Some cheaper built zappers (not necessarily cheap in price) even have a range of +/- 20 percent. This is actually acceptable according to Dr. Clark as she claims that any frequency between 10 Hz and 500,000 Hz will produce beneficial results. To an extent, this may be true, however, better zappers such as ParaZapper CC2 or ParaZapper UZI-3 will be more accurate than others, as close as o.25 percent.

Even better zappers such as the MY-3 series will have an accuracy of better than 0.05 percent. That is less than a half of a tenth of a percent error.

In his studies, Dr. Robert o. Becker found that only a D.C. voltage was necessary if applied directly to the area in question. He only worked with eliminating bacteria and tissue healing however.

The FDA and the Zapper

It is the position of the FDA that the zapper does not work and that anyone selling it is a con and a fraud. At the same time, the FDA has issued warnings to companies not to sell zappers because it is considered a CLASS III device, which requires FDA approval and a doctor's prescription. This puts the zapper in the same group of devices such as cardiac pacemakers and other life saving equipment, "Capable of saving lives". This is not at all surprising due to the fact that the FDA receives almost all of its funding from pharmaceutical companies. At the same time, the FDA has published papers that show that electrical pulses actually do kill microbes. In order to avoid qualifying zappers, they suggest using only high voltage pulses and only for purifying food and drinks. Their paper does, however, clearly state that lower voltages can be effective if the pulses are continued over time.

When surveyed back in 2004 through 2006, almost 95 percent of our customers report positive benefits from using **ParaZapper**.

Survey Results

From 2004 -2005 ParaZapper surveys using dual frequency.

**Percentage of users reporting positive results
(greater than 30% improvement)**

Problem	Overall Positive Results
Parasites	85.3 percent
Candida	95.2 percent
Colds	96.5 percent
Allergies	65.2 percent
IBS / Colitis	56.6 percent
Flu	93.7 percent
Fungus	67.6 percent
Lyme Disease	90 percent
Herpes	60 percent
Shingles	90 percent

Users with footpads report up to twice the success of those who do not use them

---

The **Hulda Clark Zapper** is legal and is readily available to customers in Canada and 26 EU countries (most of Europe) while the FDA continues to suppress its sale and use in the US.

Safety and Injury?

We have inquired of the FDA any information that they have in relation to injuries or difficulties caused by ParaZapper or any other zapper product. They could not provide us with any valid claims against the zapper. In fact, we made repeated FOIA inquiries in which they claimed they had no information about this.

NOTICES and WARNINGS

Always! Seek the advice of a licensed medical professional with regards to any health issue.

Warning! Do not use any zapper type product if you are wearing a pacemaker or defibrillator, This is not a problem with the zapper but is actually an issue with the sensing circuitry of the pacemaker or defribulator. Also, it is recommended not to use a zapper if you are pregnant or might become pregnant. There is not a known problem but there is an obligation to protect the unborn child.

Warning! Do not use any zapper when connected to a wall transformer, battery charger, or wall adapter, unless the adapter is rated for medical patient usage. This could cause electrocution.

Notice: The ParaZapper™ is offered only as an experimental device for killing microbes in water and for that purpose, FDA approval is not needed and has not been applied for. Therefore, the Food and Drug Administration have not evaluated these products. It is provided for informational use only and is not intended to replace the advice of a licensed professional. These products are not intended to diagnose, treat, cure or prevent any disease, disorder, pain, injury, deformity, or physical or mental condition. This notice is required by the Federal Food, Drug and Cosmetic Act.

Important: It has been advised not to use any zapper if the user has had surgical prosthesis of metallic material installed that is stainless steel. **This is because of the possibility of metal ions leaching from the prosthesis, causing irritation.** Irritation of tissue near prosthesis is a possibility. The main concern is with stainless steel prosthesis. We have not had any significant issues reported with Titanium implants. Reported problems include pain and swelling.

See: **Zapping and Metals**

Warning! Do not use any zapper when being treated with strong and long-term anticoagulant therapy, except under the direction of a licensed physician. A reduction of platelets may occur. The zapper is believed by some to reduce the clumping of blood cells and reduces the viscosity of blood.

Warning! Epileptics and others with similar conditions should consult with their physician before using the zapper or similar electronic stimulus. Epileptics are generally considered to be sensitive to repeated pulses.

Warning! Do not connect to any devices powered by an AC outlet or a wall outlet.

Warning! Keep snap leads and other small parts away from small children as they may present a choking or strangulation hazard. These also may contain nickel, chromium, lead, and other toxic metals that present a hazard if swallowed.

Notice! Any suggestions made in this book are based on user experience and are not a prescription for medication or cure. It is up to each user to determine his or her treatment routine. As these products are offered as research devices, the user is requested to provide Para Systems, Inc. with results and experience information.

Notice: Do not place copper or nickel plated items in the mouth. This may cause metal poisoning.

Warning: Do not place electrodes across the heart or near the chest area. While it is not likely at the low currents provided, heart stoppage should be considered a risk. This is particularly true with individuals who may have irritable hearts.

12

Warning: Keep the electrodes away from the eyes. While not documented, it is thought that the electrical pulses might overwhelm either the rods or cones in the eyes or possibly the optic nerve.

Reported Problems

Users have reported the following problems in our last 4000 sales (as of 2005):
Please note: Not all problems are reported. **See Reactions, About die-off.**
Rash or inflammation appearing on the arms or hands or other areas: 10 cases
Rash or inflammation appearing on the face: 1 case
Possible causes: sensitivity to saline solution, prolonged dampness on the skin.
Normally disappears within a couple of days.

Wrist burns from using wristbands: 7
Diarrhea: 9
Possibly did not replenish intestinal inhabitants.
Usually leaves after a couple of days.

Itching or crawling under skin: 7
Possibly parasites not yet killed, problem disappeared with continued zapping.
Heart racing with fibromyalgia. 1
Heart racing / other 1
Irritation, pain, or swelling around stainless steel prosthesis: 4

Questions and Answers

Does it work?

I have tested its functionality "in vitro" and it does work. It has killed or stopped mold growing on bread that has been left out. It has killed or stopped growth of bacteria on meat that has been left out. It kills or prevents growth in protozoa and bacteria. Identical prepared solutions of growth media we placed side by side, with one being zapped 4 times a day and the other not being zapped. The zapped media remained clear while the other control media became cloudy with bacteria and protozoa. If you have any doubts, try it yourself.

Additionally, we observed protozoa under the microscope while they were being zapped. When the zapper was turned on, they immediately moved away from the positive terminal and toward the neutral/negative. After a while, the protozoa began to swell, losing their coordination and their ability to move about. Soon afterward, internal organelles within the protozoa became disorganized and eventually the outer cell membrane ruptured causing the protozoa to explode dumping its cellular contents into the fluid around it. All species of protozoa tested this way followed the same path of destruction. You can see an actual video of this.

In a trial test, Dr. Robert Thiel found that 97% of participants reported improvement within 45 days (ANMA Monitor 24):5-9.1998). http://www.paradevices.com/thiel.html

Please note that there is one seller of a competing product has posted distorted and poorly informed interpretation of this study. He does this in order to boost his own sales.

Ok, so it works, how is it used?

ParaZapper™ is easy to use. In its simplest form, it is provided with two hand held "paddles" one of which is held in each hand while the zapper is running. The use of a 4 point contact, both hands and feet, is likely to produce far better results. The power switch is turned on and zapping continues for 7 or more minutes. The unit is then turned off for 20 to 30 minutes. The process is then repeated for another 7 minutes. The cycle is then repeated a third time. It is recommended that the process be repeated once daily to prevent new infection. Better results are obtained with 14-min. **split frequency** zaps. Even better is using multiple frequencies over a continuous period of time.

Why 7 minutes on?

Microscopic observation indicates that 7 minutes on is the minimum time needed to kill many microbes. Less than 7 minutes per frequency just is not as effective.

Why 20 minutes off?

It is felt that many parasites carry parasite eggs, fungal spores, and other resistant protected microbes inside of them. The 20 minute period allows these spores / eggs to hatch out of their protected state and become susceptible to zapping. When using multiple frequencies (8 frequencies or more), the wait period becomes less important but can still be helpful.

Alternating Leads

As the lead with the green wire is essentially at ground or zero volt potential, it is basically ineffective at killing parasites. If the same arm is always used for the green lead, problems may occur. For this reason along, it is best to swap the red and green leads each time that the zapper is used.

Importance of polarity

The active signal comes from the positive or red wires. The green wire is neutral or ground and the areas near where it touches offer protection to microbes or parasites. If you need to place a lead over an area of interest or need, then place a red wire lead there. For example, when I feel like I may be getting a sore throat, I will place the red wire paddle over my throat area.

Inverted polarity and the Beck Zapper

There are some zappers that use inverted polarity such as the Bob Beck Blood Electrifier (BBBE) and the ParaZapper MX-2. These conform to the rule in that one electrode or the other is always positive. The signal is alternately inverted so that it may appear to be negative but in reality, it is the other electrode that has the positive signal applied. The Beck zapper has the disadvantage of not being positive offset.

Rife machines

Other devices such as true Rife machines utilize sine waves to modulate a carrier wave. The purpose is basically the same, to create a harmonic frequency equal to the Mortal Oscillatory Rate (MOR) which brings about the destruction of the organism. The problem with Rife machines is that it is necessary to know the organism that causes the problem and the frequency for that particular organism, whereas the zapper produces a wide range of frequencies at the same time. The Rife machine is very specialized and requires more skill and knowledge to properly operate.

Foot Bath Detoxification

The principle of this device is to remove metals and other toxins through the feet. A big problem is that in the past some sellers have added things to the foot bath to make it appear more dramatic. As a result, it has often been declared to be a scam.

However, it can and does work because the feet were designed as an excretory organ and an absorption organ as well. This is why the feet can easily become smelly.

Is the zapper safe to use?

Yes, however, you should not use any zapper or similar product if you are wearing a pacemaker. This is not a problem with the zapper but is instead a problem of the pacemaker / defibrillator. Also the zapper should not be used on a person who is or might be pregnant. There has never been a reported problem but it is good to err on the side of caution. Do not over use or abuse the zapper. Always seek the advice of a medical professional with regards to any health issue. Zappers of various designs have been used over 25 years with no dangerous side effects. You might feel a slight electrical tingle.

Are there any side effects?

Yes, **extensive zapping with some of the better zappers can kill beneficial bacteria in the intestines or elsewhere,** causing intestinal irregularity. Eating kefir, yogurt, or drinking a smoothie or buttermilk (regular or low fat milk will work but not as well) can restore intestinal regularity. There are also helpful nutritional supplements.

Also, some users may experience effects generally referred to as die-off. When a lot of microbes are killed in the body, the dead microbes will release a lot of toxins that can make you feel sick for a while.

Some users complain of wrist strap burns.

Improper zapping can cause problems.
See: Reported problems under Notices and Warnings

Is one brand or model of zapper better?

Due to the Current control in ParaZapper™, it provides a stronger output than many other zappers and uses less power from the battery thereby making the battery last longer. This is achieved through the use of better, more expensive components.

Some other zappers cycle on and off automatically.

Most ParaZapper™ models do not automatically cycle because we feel that the user should have control over the length of time that they zap due to individual differences. Otherwise, most zappers perform the same function, which is outputting a 9-volt square wave at some frequency.

A lower frequency of 2.5 kHz is probably better for some parasites as discussed above. ParaZapper™ CC2, UZI-3, and MY-3 definitely produce better results according to user feedback.

What happens to the parasites that are killed?

The bodies defense mechanisms have to clean up and remove the killed organisms or the debris from their destruction. For this reason, you may feel fatigued throughout the day following zapping. Headaches, itching, nervousness, and diarrhea are possible. This is often referred to as die-off. Drink plenty of water (3-4 glasses) and take it easy for the day.

Does it matter which paddle goes in which hand?

No, not really, but we do recommend that you swap paddles each time that you turn the zapper on. Red lead in the right hand one time, red lead in the left hand the next time that the zapper is turned on. Balance is good.

Do I have to hold the paddles in my hands?

No, some people place them under their feet. Others have sat on them. This should be acceptable as long as the paddles do not touch. The green paddle protects parasites where it touches and should be in a hand or under foot. It is important to note here that 4 point contact (with 3 points positive) will generally produce better results.

Should I feel anything while zapping?

While some individuals may feel a slight tingling, it is not necessary and certainly not a problem. If you use the 4 point contact system with 3 points positive, the green wire lead will definitely feel stronger.

Can I get electrocuted?

No, not from ParaZapper™ as the current output of ParaZapper™ is limited by built in resistance. Also, the output voltage is less than 11 volts, which is basically safe in itself.

Warning: do not connect to any devices powered by an AC outlet.

I did not get good results, what can I do?

There are 3 main reasons that users report poor results.

! The first is from not using the saline solution to increase conductivity. **Use the strong saline**, it is important. **NOTE:** Saline for the eyes is not strong enough!

! The second group reporting poor results were those who use only the wrist straps. The **copper paddles are at least 4 - 5 times as effective.**

! The third group reporting problems did not zap for a sufficient period of time. They often report zapping for only 5 or 6 minute sessions, only 1 or 2 zappings per session, or less than 1 hour.

Always zap at least 7 minutes on each frequency, use both, and zap 3 or 4 times with 20 rests minutes between. Copper footpads combined with the copper hand paddles provide the best results.

Remember: 7 minutes minimum on each frequency, repeat each frequency 3 times minimum, rest 20 minutes between.

Recommended:

While it is far better to use multiple frequencies, **when using dual frequencies** it is suggested to zap 7 minutes at 30,000 Hz, 7 minutes at 2500 Hz with 16 to 30 minute breaks between. A 20-minute break appears to be ideal but not critical. Repeat 3 times minimum. Some users report zapping for over an hour continuous without any problems but this does appear to be less effective than including the breaks, less than 1 hour does not work well.

However, when using a multi-frequency zapper or a multi-frequency mode, 1 hour continuous or longer works better.

Many users also **wrap the copper tubes with a wet paper towel (using a solution of sea salt)** to keep the copper off of their skin. This also reduces resistance and applies a more powerful signal to you. This is recommended. Also, cover the footpads with a damp paper towel to keep the copper off of your feet.

Do not drink the salt water solution!
Use the sea salt and towels.

Cleaning my ParaZapper™

The case may be cleaned with a damp cloth and mild soap. Do not soak or allow any water to get inside!

The paddles may be cleaned and shined with steel wool. You may also use Brasso for cleaning making sure to clean all of the Brasso and residue off when finished.

Copper footpads used with the copper paddles produce even better results. Do not use footpads with wrist straps. This may cause skin burns.

See electrode oxidation.

The LED on my zapper comes on; can I be sure it is working? **Yes, this is easy. Find an AM radio (FM will not work for this) and turn it on. With the zapper turned on, turn the tuning dial until a whine or squeal is heard.**

Turn the zapper off and on several times. If the whine stops and starts with the off and on, then the zapper is working. Note: the LED is powered by the zapper outputs so if the LED is on, then the zapper is probably working.

Most recent model ParaZapper products have a tri-color LED that will show yellow or amber if the zapper is working properly. There are further instructions on testing your zapper in another section of this manual.

Health and Illnesses

Please report your success or failure to us when zapping various problems. If possible, keep notes in the back of this book. See Complimentary Therapies.

Abscesses

The zapper has been frequently used to help reduce the pain and swelling of boils, carbuncles, and other infections under the skin, including abscessed teeth and jaws. The positive electrode is normally placed over the affected area and normal zapping timing is followed. A tooth abscess can often be helped in this way but be sure to see a dentist as a lingering infection may cause serious problems. The red wire paddle may be held in place on the outside of the jaw directly over the affected area and green wire paddle should be held in the opposite hand.

Acne

Zappers have been reported to help reduce acne problems by many users. Note that zapping will not remove all causes of acne but may help control skin pathogens.

Allergies

Since many allergies are the result of or are aggravated by certain parasites, relief can be obtained by removing these parasites or microbes. This may take weeks or months to notice in some cases. When surveyed, over 60 percent of responding users reported noticeable positive results. When using augmentation footpads, this improves to over 75 percent, demonstrating the importance of using the 4 point contact system. Keep zapping. Molds and fungi can definitely be a problem.

Alzheimer's

While customers have not been surveyed for improvement in this condition, there have been anecdotal reports of improvement. It is important not to apply the electrodes directly to the head or neck area, especially not near the eyes. There are some who believe that the 5.00 kHz and 10.0 kHz that is present in some models may be helpful. This has not been supported by any scientific studies.

Arthritis and arthritic symptoms

Many zapper users claim to get relief from arthritis and arthritic symptoms. This is a good indication that many cases of arthritis are either caused by or aggravated by microbial infections. Allow several weeks to notice significant results. When surveyed, over 60 percent of responding users reported noticeable positive results. Of those who use the footpads, 84 percent report positive results, again supporting the added benefit of the 4 point contact system with 3 points positive.

Asthma

Medical literature indicates that asthma and similar conditions may be caused and aggravated by parasites. It may be necessary to zap for 45 to 60 days for full effect. Keep zapping daily. When surveyed, over 40 percent to 60 percent of responding users reported noticeable positive results.

Athletes foot and fungus

We have had several users report relief from fungal problems using ParaZapper™. The use of footpads is very helpful. In one particular case of serious fungus on one foot (due to wearing wet socks for 12 hours), the user zapped a full cycle of 7-7-20-7-7-20-7-7 and then zapped a single zapping every 8 hours using copper foot pads in the substitution mode for 5 days. Itching and burning stopped within a half-hour. The fungus was killed within 3 days. It took about 8 days for the cracked skin to heal. Also, instead of saline, Use 1 tsp. sea salt, 1 tsp. Epsom Salts, and ½ tsp. baking soda in 4 - 6 oz. warm water. The user reported

that using either Tinactin nor Clotrimazol DID NOT have sufficient results in reducing this particular infection over a period of 14 days.
Infection may recur due to spores left behind.
Frequent or chronic fungus infections may be a sign of Candida infection.

Candida

Many users find that Candida needs longer and more frequent zapping sessions and responds better to 2500 Hz but both frequencies are beneficial. Also, on advanced zappers, try 727,787,802,880. Also check the Rife CAFL. Some of the more advanced zappers like the MY-3 have specific modes for both fungal microbes and for Candida microbes. These frequencies are not scientifically proven but may users feel that they have good results.

Suggestion: For best results, for the first two or three days, eat nothing and drink only water for a few hours before zapping (24 hours is best). This will leave the yeast in a weakened state. Zapping four or more zappings in a session with the 4 point contact system, augmentation mode, may produce better results. After zapping, flush with Epsom salts to clean the intestines. Following this, take some probiotics along with a small amount of kefir, yogurt or a smoothie Take about 1 tablespoon full of good yogurt every 15 to 20 minutes for the next hour or so. It may be beneficial to wait a 3 or 4 of hours before eating anything else. Following this avoid sugars, alcohol, and carbohydrates as the Candida yeast thrives on these. Normalize your pH.
Candida reproduces at a fast rate and if you kill 99%, it can recover in a few hours. **See: Overdrive**
Read: "The Yeast Connection" by William. G. Crook M.D.
Note: Not replenishing normal intestinal bacteria after zapping can increase your chance of getting Candida.
Note: Candida can form threads through the intestine walls, which leaves pores when it dies from zapping.

Cancer

The original purpose of the zapper was to treat cancer by removing the parasites that cause or aid cancer or inhibit the immune system. The **National Cancer Institute** and the **National Institute of Health** now recognize that certain parasites do cause cancer. The list includes Liver Flukes.

I do not claim that the zapper can cure cancer but there are numerous reports of cancer sufferers who use the zapper that go into remission. My personal experience is that the zapper quickly and efficiently dispatched my basal cell cancer which had spread to several locations on my head.

Surviving cancer requires a healthy immune system. It may be smart to not take any medication, chemotherapy, radiation, etc. that might negatively affect the immune system. Because the zapper helps remove parasites, it should help cancer sufferers by allowing the immune system to concentrate on cancer. As an experiment, hold the red wire paddle oever the area of concern with one hand and the green wire paddle held in the opposite hand away from the body. Using the copper pads at the same time as a 4 point contact system, 3 points positive, will force additional energy through the torso.

See http://www.paradevices.com/cancer.html

Alternatves: Essiac Tea, Cancell, Procell, PawPaw, Cesium.

Chronic fatigue syndrome and malaise

One of the most frequent causes of fatigue is a parasite infestation. Many people with this condition have a virus such as Epstein Barr or possibly some other microbes. Because a heavy parasitic infection can steal nutrition and rob you of your energy, this can be a major cause of chronic fatigue. Some users report improved energy levels within a few days of frequent heavy zapping. For best results, use the copper footpads in the augmentation mode.

Chronic fatigue has also been associated with certain other virus infections.

Colitis

As Colitis, IBS (Irritable Bowel Syndrome), and frequent diarrhea are often the result of certain parasites in the intestine, relief can be obtained by removing these parasites.

Note: The original zapper is not as effective at relieving IBS as it does not have the 2500 Hz frequency. Multi-frequency works much better.

Copper footpads are also very helpful. Customers who use the copper footpads along with the copper handles report better results than those who do not.

See: user suggestions: start with caution.

Colds and Flu

Colds and Influenza are caused by viri and there are several of them. They all seem to be sensitive to zapping. Many zapper users report stopping colds and flu **by zapping as soon as symptoms appear**. This has been verified by at least 1 limited university study group. It is speculated that this may be a result of increased immune system activity. For best results, zap at the earliest signs of cold of flu. Use the copper footpads in the augmentation mode if possible. Also zap again if symptoms persist or return. Multi-frequency modes appear to be more effective than just 1 or 2 frequencies but in my experience, there have been times that I stopped oncoming flu with just a single one hour session with a single frequency zapper.

NOTE: Once cold or flu has been established, zapping often does not have as significant of an effect. It is suspected that the reason for this is that the virus has already infected a large number of cells.

When surveyed, over 94 percent of responding users reported noticeable improvement in their colds from zapping. About 47 percent of respondents showed 75 percent improvement or better. Customers who use the **copper footpads** along with the copper handles report better results. 64 percent of footpad users reported an improvement of 75 percent or better.

Also, when surveyed, 92 percent of flu sufferers who responded reported noticeable improvement in their condition, of those who used footpads with copper paddles, 91 percent reported better than 50 percent improvement.

Diarrhea

Frequent diarrhea may be the result of certain parasites or microbes in the intestine. Relief can be obtained by removing these parasites.

Note: The original zapper is not as effective at relieving IBS as it does not have the 2500 Hz frequency. See **Colitis** above.

Dental problems

While not a substitute for good dental care, use of **ParaZapper**™ can reduce the pain and swelling of abscesses as well as ease inflammation of the gums.
As tooth ache and gum disease can lead to serious and even life threatening problems, consult with your dentist. For easing the discomfort and assisting treatment, hold the copper paddle with the red lead (wrapped in a saline dampened paper towel) against the outside of the affected area of the jaw. Do not place the paddles or other metal parts inside of the mouth.
When surveyed, 70 percent of respondents reported noticeable improvement in their dental problem when zapping. These effects included the reduction of pain and the reduction of swelling.

Fibromyalgia

28

We have several customers using **ParaZapper**™ for Fibromyalgia. One reported problems with their heart racing when zapping. The others have not had this problem.

When surveyed, a limited number of fibromyalgia sufferers responded but of those, almost all reported at least 50 percent improvement in their condition.

Fiery Dragon

This parasitic worm comes from Asia but does show up in North America and the female burrows under the skin laying its eggs. This causes an allergic reaction with burning under the skin. When zapping is started, this burning may intensify due to movement of the worms but should subside within a couple of days. Placing the copper paddle with the red wire directly over the affected area may produce maximum effectiveness.

Flatulence, bloating, and gas

These are frequent symptoms of intestinal parasitic infestations.

Note: The original zapper is not as effective at relieving these problems as it does not have the 2500 Hz frequency. See **Colitis** above.

Food allergy

Many food allergies can be caused by the presence of parasites. Over 70 percent of customers report positive results from using the zapper for allergies.

Hepatitis B and C

ParaZapper is being used by Hepatitis B and Hepatitis C infected individuals to help keep their infections in check. Basically, the more you zap the better the results. If you have Hep-B or Hep-C, please keep us posted as to your results. **ParaZapper AV** was available for those who were near terminal or those who have a high viral load. Unfortunately, the FDA has forced us to remove this product from the market. The ParaZapper MY-3 has frequencies that have been reported to be helpful in killing the HEP-B and HEP-C virus. These frequencies are not proven. Statistical results were not collected for HEP-B or HEP-C.

Herpes

A number of customers claim to use ParaZapper™ with good success in reducing herpes outbreaks.

When surveyed, 85 percent of responding users reported noticeable improvement and those who used the footpads reported better success.

As a child and in my adult years, I had frequent oral herpes but this went away after I started using the zapper and I have never had another outbreak.

HIV / AIDS

Previously, ParaZapper was being used by HIV infected individuals and aids patients in clinics to fight secondary infections as well as an aid in fighting the HIV virus. There are not any current clinical studies underway with ParaZapper and AIDS patients at universities.

Due to the dynamics of HIV, it may be beneficial to zap very frequently. HIV reproduces at a rate of nearly 100 million virus units per day. This replication is fairly constant so that the more frequent the zapping, the better the results should be. Some users will use the advanced zappers like the MY-3 for 12 to 20 hours a day. Also, the use of **copper footpads** in the **augmentation mode** should be very helpful. Please let us know anything that you may find helpful.

See Overdrive

ParaZapper™AV is not available for HIV positive individuals with high viral loads. There have been positive results in limits study trials with ParaZapper™MFC. These results are not published.

IBS and colitis

As Colitis and IBS (Irritable Bowel Syndrome) are frequently the result of certain parasites in the intestine, relief can be obtained by removing these parasites. Over 80 percent of ParaZapper users reported noticeable positive results in improvement of their IBS / colitis.

Note: The original zapper is not as effective at relieving IBS as it does not have the 2500 Hz frequency.

Customers who use the **copper footpads** along with the copper handles report twice as good of results.

Intestinal worms

ParaZapper is reported by some users to be effective in the treatment of intestinal worm infestations but there have not been any studies to support this. We get frequent calls telling us of what has been passed after zapping. These range from spaghetti masses (round worms) to tapeworms (strings of rice). The **copper footpads** are very beneficial. It seems that some users will have great success when zapping intestinal parasites while others do not seem to succeed. It could be the result in the difference of frequency between units or it may be due to individual differences in conductivity, etc.

Joint aches, joint problems and joint pains

Some joint problems can be caused or aggravated by the presence of parasites or microbes. Some individuals report relief after a few days of zapping. Others may require a longer time. See: Arthritis.

Lyme Disease

Several users have reported good results with reducing the effects of Lyme disease. The 2500 Hz seems to reach better. Multi-frequency seem to work even better.

Users have found that they may get better results with zappers such as the MY-3 that has specific frequencies and modes.

My personal experience with Lyme (actually, because of my location, it was referred to as STARI, southern tick associated rash illness) was in 2004 where I was bitten on the abdomen by a tick and shortly after developed the characteristic bulls-eye pattern. I immediately started zapping with the ParaZapper MX in the 2500 Hz mode for several days. I have not noticed any symptoms developing since them.

Malaise

Malaise or a general lack of energy can be due to a heavy parasitic infection, which can steal nutrition and rob you of your energy.

Malaria

Considering the fact that Malaria is protozoan in nature (Plasmodium), ParaZapper should be particularly effective in treatment of this illness. This has not been proven or disproved. **See Overdrive.**

Morgellons

The effect of the zapper on Morgellons sufferers has not been documented. Several individuals with Morgellons have called and expressed very good results from zapping but statistics have not been collected. Most users seem to prefer the 2500 Hz mode. Tub zapping seems to be helpful.

MS or multiple sclerosis

A number of individuals are successfully using ParaZapper to fight the effects of MS. The more powerful **ParaZapper™** products are preferred and seem to produce better results. Recovery does take months to years. The most important result is stopping the progression of MS. ParaZapper was featured in an article in the NEW PATHWAYS Journal of the MSRC.

Parkinson's Disease

Users have reported good results in reducing the effects of Parkinson's disease with ParaZapper™CCa and with ParaZapper™MX . It is felt by some the the 5 kHz and 10 kHz modes in the ParaZapper MY-3 is helpful. Again, there are not any scientific studies to support this.

Salmonella

While only a few users have reported having Salmonella poisoning, All have reported better than 75 percent improvement after zapping with ParaZapper when using paddles and pads.

Sepsis and Septicemia

While any of the ParaZapper™ products should be beneficial in killing bacteria multi-frequency is recommended for extreme cases as it provides a broad range of frequencies. Many individuals have reported good results in reducing bacterial infections.

Constant zapping at 2500 Hz for 14 minutes on, 20 minutes off may be beneficial in reducing severe bacterial infections. Use of footpads may significantly increase results.

My personal experience with this was from a deep wound infection in my left foot. The foot was swollen to twice its normal size and showed extensive reddening and red streaks. Also had a fever present and lost most feeling in the foot. It took several hours of zapping with the ParaZapper MY-3 in the MX mode to stop this infection.

Shingles

Several users have reported good results in controlling shingles. It would likely be most effective if used early in the outbreak. There has not been any study reported to date.

Sinus problems

For sinus problems, the copper paddle with the red wire can be held against the face but should be below the cheek bone to avoid possible eye injury. I usually place the red wire paddle against the inflamed side of the nose and hold the green wire paddle in the opposite hand.

Skin problems, itching, and rashes

Some of these problems may be caused by parasites. If itching is caused by parasite, the itching may increase during the initial stages of zapping. If itching occurs in the legs or lower torso, use copper footpads for greater effectiveness.

Tropical diseases

There are numerous tropical illnesses caused by parasites. ParaZapper™ should be effective against any of these. In the case of illnesses that exhibit rapid propagation such as Malaria, it may be necessary to use the zapper frequently. The most likely time for usage is at the time when the parasites break out of their intracellular condition and invade the circulatory system in mass.

The ParaZapper™ has been reportedly used successfully against Chlamydia, Giardia, Candida, pinworms (Enterobius), whipworms (Trichuris), and flukes (Fasciola) among others, according to customer reports.

West Nile Virus

We have recently had reports from customers who used ParaZapper™ to treat the after effects of West Nile Virus (medically diagnosed as West Nile) with success. We have not heard from anyone who has used ParaZapper™ to stop West Nile in the early stages because the disease was already in progress before the purchase. If symptoms of West Nile Virus appear, contact your nearest medical professional immediately. As West Nile

34

is a virus, zapping as if it were a cold or flu should produce beneficial results.

SARS, Bird Flu, Marburg, Ebola, Polio, Measles
While we have not tested ParaZapper™ on these, good results should be possible when using ParaZapper™ as you would for influenza. In life threatening cases do not hesitate to use extensively. In such cases, ParaZapper™ is not likely to cause any harm. Please report any positive or negative results to us.

Other uses

Water treatment

The zapper can be used to help clean up biologically contaminated water in ponds, buckets, holding tanks, etc.

Using ParaZapper for water purification is easy and simple. The device is designed to pass electrical pulses through the water and kill free living parasitic and pathogenic organisms. It has been consistently demonstrated that this is an effective means of eliminating microorganisms from water. It is not recommended for use on drinking or potable water but in an emergency the benefits may outweigh the negative aspects. Using silver electrodes will eliminate most of the negative issues.

The process is simple and involves placing either copper electrodes or silver electrodes into 1 to 5 gallons of clean filtered and decanted water. The filtration is recommended to remove scale and sediment that can interfere with the electrical signal. A coffee filter is usually sufficient for the first stage filtration. This will remove larger particles and possibly some larger parasites. Following this, it is beneficial to allow any cloudy water to stand for 24 hours to allow any sediment to settle out. The clear water can be siphoned from the top into a fresh clean bucket for zapping.

Once the electrodes are in place, the power to the unit is turned on and kept on for 30 minutes. Stir the water every 5 minutes while the electricity is applied. After 30 minutes, the power is then turned off, and the electrodes are removed. The water is left undisturbed for 30 minutes and then the process is repeated for a total of 3 times.

At this point, the water should be free of most organisms. However, some organisms such as Cryptosporidium require a stimulus to become active. Add 3 drops of sodium hypochlorite (chlorine bleach) for each gallon of water, stir, and wait 2 to 5 hours to allow these parasites to hatch out. Repeat the zapping process. **ParaZapper is not sold or intended for drinking water purification**.

The copper electrodes are not recommended for use with water for human consumption or for animals and silver rods (minimum 99.9 %) three-nines pure should be used for treating all water used for emergency human consumption. Use of silver with less than 99.9 percent pure may contain undesirable contaminants.

Notice, the use of sodium hypochlorite can introduce small amounts of copper chloride or silver chloride into the water. These amounts are within acceptable levels considering the use of correct timing. Do not leave the electrodes in the water.

Food Purification

Note: Of all the speaker boxes (food zappicators) that I have tested, none have worked. The speaker box is basically a box with a "north pole" speaker inside of it as the orientation of the speaker is important. The speaker is only connected to the positive output of the ParaZapper. The ParaZapper is turned on for 7 minutes or more, then off for 20 minutes. This process is repeated 3 times minimum to successfully kill parasites in the food. **Testing:** I tested by placing ice cube trays filled with infusoria on top of the speaker boxes which were activated for up to several hours but the microbes were not killed in significant numbers. **ParaZapper is not sold or intended for food purification.**

Aquarium Parasites

Use caution when treating any aquarium that contain invertebrate specimens such as shrimp, crabs, starfish, sponges, etc, as the effect of zapping on these organisms is not known.

Hanging plate electrodes at each end of the aquarium and zapping according to normal routines can easily treat fish with diseases such as fungi, ich, and bacterial infections. Avoid using copper electrodes as the copper can kill many plants. Titanium may be the safest electrode for this but it is expensive and untested.

37

Stainless steel may also be acceptable in the aquarium but the overall effect is not known. It is known that stainless steel contains nickel, chromium, and cobalt, all of which are toxic.

How to Zap single and dual frequency

Using Saline Solution for Zapping

We send out a survey to our customers after a couple of months using the zapper. The important information that we have received is that those who use the saline solution report better results. This would be true for any and all zappers, not just ours.

NOTE: **Saline for the eyes is not strong enough!**

User suggestions: Start with caution

When using a single frequency or dual frequency zapper, it is wise to start with caution. Zap 1 session of 7-20-7-20-7 (7 on, 20 off) for the first 1 or 2 days at 30 kHz. Do not rush unless you are in critical condition.

Accelerate

After the first couple of days increase the zapping time to 14 minutes (should be 7 min at 30 kHz then 7 min at 2.5 kHz) such as 14-20-14-20-14 (7-7-20-7-7-20-7-7). Keep this for 2 or 3 days.

Shift to high gear

The next step up is 4 cycles per session. 14-20-14-20-14-20-14. Better is 7 min At 30 kHz and 7 min at 2.5 kHz. This method is beneficial to individuals with more resistant parasites.

Overdrive

Individuals with extreme conditions may want to try the 4 cycle sessions twice a day. Even better results may be obtained with additional single zappings every 4 hours or even hourly. This can be beneficial in rapidly propagating viruses, protozoa, and bacteria. This may also be beneficial in certain parasites that spend part of their time intracellularly. These parasites are probably protected in their intracellular mode and reachable only when they break out of the cell.

39

Continuous Zapping

Continuous zapping of less than one hour does not work very well for single or dual frequency. Over 1 hour is not as good as 4 cycles of 14 / 20 unless multiple frequencies are used.

Tub Zapping

The copper pads and red wire paddles are placed in the tub along with ½ cup Epsom salts and ½ cup sea salt. The user gets in and holds the green wire electrode in one hand or the other and runs the zapper through its normal cycle.

How to Zap multiple frequency

Since most multiple frequency zappers have instructions for the timing of application, it is best to follow their instructions or suggestions.

It is important to continue zapping until all the frequencies in a mode have been applied as those frequencies are designed to work together and to support each other.

Using multiple frequencies can also help eliminate the need to repeat applications when continuous zapping is used but there may be situations where it may be beneficial to stop at the end of a continuous mode, wait for a while, 20 minutes to an hour, and then run the frequencies a second time.

Overdrive

In certain situations, is is a good thing to just keep using a zapper for many hours a day, every day. People who have successfully tried zappers for things such as reduction of viral loads and increasing T-cells, may often use the zapper for 16 hours a day or more. Just remember to keep the paper towels or cloth wet, not just damp, and change them every few hours.

Be careful of possible skin burns when using for long periods of time. Allowing the paper towels to dry can increase this risk.

Using the ParaZapper™ parasite zapper

How frequently should I Zap?

The frequency and amount of zapping depends on the individual, his or her condition, and the type and extent of illness being treated. Starting out, most users zap 1 session (a session is 3-4 cycles of zapping) a day. This is normal for the first 2 to 3 days. This may be followed with more aggressive zapping for 2 weeks up to 6 weeks. Afterward, users may zap from once a day to a couple of times per week. Once parasites have been reduced or eliminated, the need to zap is reduced. Zapping should not be discontinued though as it helps prevent re-infection.

How long should each zapping session last?

Recommended normal zapping is for 7-14 minutes with a 20 minute resting stage in between. This **must be repeated 3-4 times for each session**. In extreme cases, some individuals choose to zap continuously for a full hour. There is probably no harm in this, but it is probably not as good as 4 cycles of 14 minutes. It is important to eat some Kefir, yogurt, or drink a smoothie after zapping.

It is **recommended** that **you zap at least 3 or 4 times per session** as some users have reported problems with zapping less than this. **4 works better.**

Can I eat or drink while zapping?

Always **wash your hands after handling copper (or any other metals)** before eating. For the first few days, it is recommended to eat only fresh raw fruits and vegetables for 2 hours before and wait 2-3 hours after zapping. While not critical, it is beneficial. **See Candida.** It is also very important to **drink 2-3 glasses of water** before, during, and after zapping sessions. It is needed to remove toxins and waste byproducts.

Take some kefir, yogurt, drink a smoothie or some buttermilk as soon as you are finished with your zapping session. Actually, eating a spoonful of kefir or yogurt several times a day is good.

42

It is also recommended that you do not take any drugs (unless prescribed by a physician) or herbal remedies for 2 hours before and 6 hours after zapping for the first 3 or 4 days. This is due to temporary increased absorption.

Why should I use salt water and paper napkins?

The saline solution **reduces the resistance of the skin surface** and **allows a stronger signal to reach your parasites.** The sponge pads or even paper napkins help keep the copper off of your skin and to hold moisture. The paper napkins do not need to be dripping wet.

How can I make the saline solution?

Sea salt is preferred and is available from most grocers. Mix one teaspoonful with a small cup of warm water. Soak a couple of thin sections of natural sponge or paper napkins in this solution. Wrap around the copper paddles or place under the metal of the wristbands.

Included with your **ParaZapper**™ is a small package of sea salt and two half paper towels to try this.

NOTE: Saline for the eyes is not strong enough!

Important

Always remove the paper towel material and discard. The bluish green tint contains copper chloride, which is toxic to humans and animals.

The paper towels supplied are Scott Rags and are available at Home Depot, Lowes, Ace Hardware, and other home improvement stores. Use only white towels.

Testing your zapper

Most ParaZappers have A Tri-color status LED that will normally show yellow or amber color (a mixture of Red and Green combined). If the red wire and green wire electrodes are shorted together, this LED should turn red or almost red.

This means that the output is functioning properly.

Checking for output current

If you check the current using a DMM or analog meter in the DC current mode, double the reading. The reason for this is that the meter integrates and averages the reading. The best way is to measure the voltage drop across a known resistor of 500 ohms. Divide the volts by the 500 ohms (I=E/R). 1 volt = 2.0 milliamps (ma). 5 volts = 10.0 milliamps (ma), etc.

Checking for output voltage

If you check the output voltage using a DMM or analog meter in the DC volts mode, double the reading. The reason for this is that the meter integrates and averages the reading. A regular Hulda Clark zapper will read about 3.5 to 4.0 volts with a full battery. ParaZapper products will read 4.6 to 4.8 volts. This is because the meter averages the square wave signal. ParaZapper UZI should read about 5.25 volts except modes 5 or 6 which will read higher due to the odd waveform.

It is strongly recommended that you complete at least three zappings for a full session. 4 zappings are better and 5 zappings have been reported to produce even better results.

Usual Steps to Zapping

1) Install battery if it is not already in.
2) Prepare your salt water solution. About a teaspoon in a cup of warm water to a tablespoonful in a 10 oz glass.
3) Connect the copper foot pads if using them (these can double your results), copper tubes (paddles), or wrist straps (least effective) to your ParaZapper. Our new straps come with a 6 inch square expansion plate for improved contact area. This is better than the old 1 inch square of the straps. (Do not use the footpads with wrist straps as serious wrist burns may occur!)
4) Place good white paper towels (or cloth) which has been soaked with salt solution around paddles or under metal plate of wrist strap. The more salt water, the better the conductivity that may be achieved.
5) If using wrist straps, put them on, one on each wrist. If these have expansion plates, put the plates on top of the wet paper towels under the metal contact of the straps (between the metal plate of the straps and the paper towel that contacts the wrist. 6) Turn on power to your ParaZapper.
7) For dual frequency zappers, Select your preferred zapping frequency 30000 Hz with button in, 2500 Hz with button out. For the advanced, more effective zappers, select the mode that you want to use.
8) If using copper tubes (paddles) pick 1 up in each hand and grasp firmly. There is no need to clench tightly though, just provide good contact area. . . You may place the paddle with the red lead against an area to treat it but use only one green lead and keep it away from the body.
9) Normally, for single or dual frequency zappers, continue for 7 minutes to 14 minutes (or as specified by the instructions if you have an advanced zapper).
10) Turn your ParaZapper off. Set down paddles or remove wrist straps.
11) For single or dual frequency zappers, wait 20 minutes.
Many advanced zappers do not need to repeat the cycle again.
12) For single or dual frequency zappers, swapping the red and green leads, repeat steps 3 through 11 above twice more for a total of 3 zappings. It is strongly recommended that your complete at least all three zappings for a full session. (4 zappings may produce even better results.)

After first zapping many individuals will either go through a phase of tiredness. Some people report immediately feeling more energetic. Others experience slight diarrhea while some experience both. Some have reported slight headaches, skin crawling, or itching skin. These conditions should not last long (a day or so). There have been rare cases of anxiety and heart racing, these appear to go away after a few days usage. Although there has never been a death or serious injury, use caution in making your decision about continuing. Parasites that have died in the body leave debris that must be removed and the body may use a lot of energy doing this. **Drink lots of water** and **get plenty of rest**. Also, the loss of intestinal inhabitants can cause diarrhea. **Replenish the intestinal flora**, drink buttermilk, eat yogurt, or take probiotics. After the first few days, many users will feel a surge in energy.

How can I get the most out of my ParaZapper™

1) Use your zapper regularly **after initial parasite treatment.** When parasites are gone, zapping a couple of times a week is usually enough. This should prevent occurrence of new problems.

1) **Use sea salt** (table salt will work) to increase effectiveness. This will reduce the effect of skin resistance.

2) **Always drink plenty of water** before, during, and after zapping. If you do not, toxic substances could possibly build up, damaging kidneys and other organs.

3) **Get plenty of rest after zapping.** Rest gives your body time to clean itself, removing dead parasites.

4) **Switch polarity each zapping session.** Use the red wire to the left arm on one session and to the right arm on the next session. **Parasites nearest the green lead probably get the least effect from zapping.** Also, this tends to reduce any ionic migration effect that may be present. **Always keep the green lead away from the body.**

5) Using both frequencies will reach more parasites but it is important to repeat each frequency 3 times minimum.

6) If you need to eat, for the first two or three days, eat only fresh fruit and raw vegetables for 2 hours before and if possible, wait 2-3 hours after zapping. This is beneficial for some.

7) It is recommended to **eat** some kefir, yogurt, drink a smoothie or some buttermilk a few minutes after zapping has finished. Acidophilus tablets are available from GNC and other health food stores but are not as good.

8) When possible, **use the copper footpads** along with the copper hand-holds for better signal transmission. **They do help.**

9) **Avoid using wrist straps** whenever possible. They are less effective. The same is true for TENS pads.

We now recommend an extended zapping program: **See: Zapping Program**

47

Zapping Program for dual frequency

From our experience, (not from any medical point of view), we use and recommend the following:

First zapping: use standard procedure, 30 kHz, 7 minutes on, 20 minutes off, 7 minutes on, 20 minutes off, 7 minutes on.
Rest, drink lots of water, eat some kefir or yogurt, or drink a smoothie or some buttermilk. Wait 24 hours before zapping again. This will avoid over-taxing your system.

Second Zapping: If you had good results from the first zapping, then repeat the same procedure again. Otherwise, try zapping 7 minutes at 30 kHz, then 7 minutes at 2.5 kHz, off for 16 - 20 minutes. Repeat 3 times.

Again, rest, drink lots of water, eat some kefir or yogurt, or drink a smoothie or buttermilk. Wait 24 hours before zapping again. This will avoid over-taxing your system.

For the third zapping, zap for 7 minutes at 30 kHz, then 7 minutes at 2.5 kHz, off for 16 - 20 minutes. Repeat 3 times, 4 times for better results, or 5 times for even better results.
Again, rest, drink lots of water, etc.

After the third day, if necessary, zap two sessions a day. **See user suggestions.**
We do not recommend zapping for more than 1 hour at a time as we doubt that there is any additional benefit. **Zapping for under 1 hour straight does not work.**

Buttons on ParaZapper™

There are 2 buttons on ParaZapper™ and ParaZapperPLUS™ and ParaZapperCCa™ models. The left button is the Power On / Off switch. The right button selects the frequency range. When the switch is out 2500 Hz is selected. 30 kHz is selected with the switch pressed in.

Using copper footpads

The augmentation copper footpads provide a more uniform signal across the body. They are designed to give a more even distribution of signal to the torso and extremities.

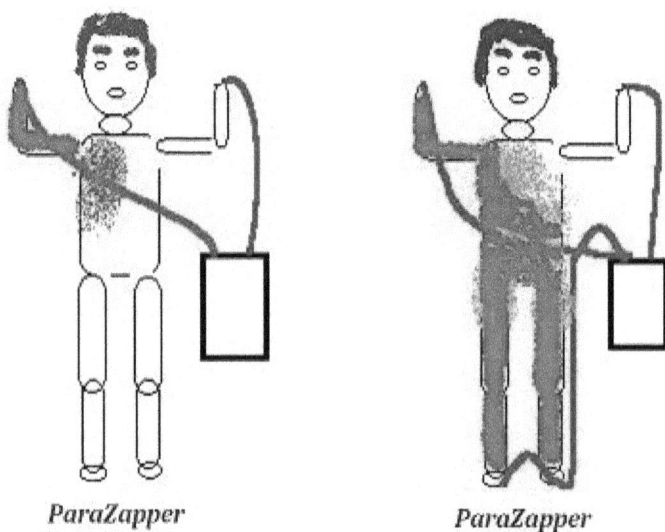

ParaZapper   ParaZapper

Place bare foot on each pad (with saline soaked napkin between foot and copper) and connect the leads to the positive (red) jack on the ParaZapper.

Stackable Plugs

The plug for the footpads is a stackable plug so that the red plug for the copper tube handle can be plugged into the plug for the footpad.

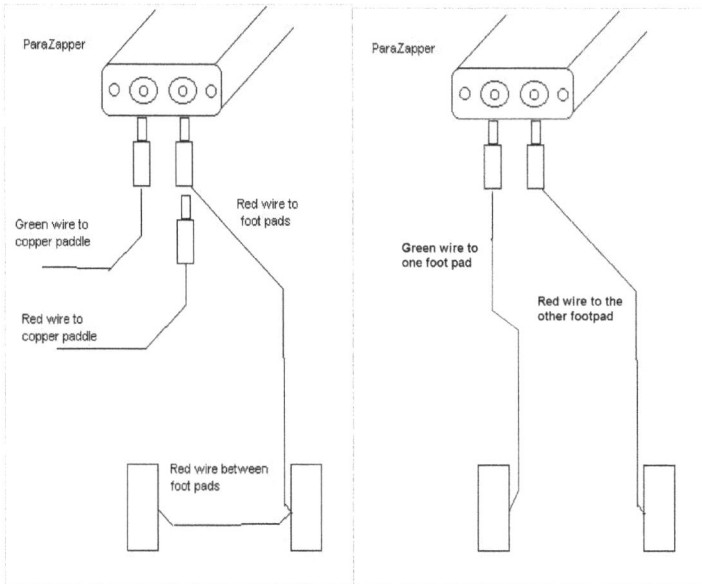

Augmentation Mode   **Substitution Mode**

50

In the substitution mode, the footpads take the place of the copper paddles. In the **Augmentation mode** the footpads provide extra signal strength to the body in addition to the copper paddles as well as multiple signal sources.

Swapping leads

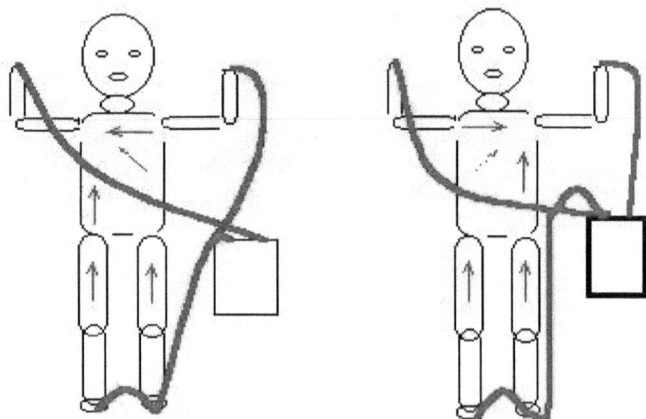

Money Back Guarantee

Check Website for current guarantee and return policies.

It is our desire for you to have a functional and useful product. If after receiving your **ParaZapper™ parasite zapper** you are not fully satisfied with the product, you may **return it to the place of purchase** in **new undamaged condition** for a full refund of the unit cost within 30 days after receipt (excluding S & H and used wrist straps). **Please call before returning.**

Used wrist straps can not be resold so there is a $10.00 US restocking charge on units returned with used wrist straps. Shipping, handling, and insurance cost are not refundable.

Additional charges may be:
Clean and polish copper paddles $7.50
Clean and polish copper pads $9.50

Ten days are not enough for full results, but are sufficient to notice a change. According to one test study on zappers, 97% of users achieved noticeable improvement within 45 days. **Call us if you do not see results in 10 days.**

Warranty

Assembled ParaZapper units are warranted against defects in materials and workmanship for a period of **5 years from date of purchase.** Not covered by this warranty, are damages resulting from misuse, abuse, neglect, accidents, or acts of god and nature. At our option, we will either repair or replace a defective unit or any part that is covered by warranty. It is the customer's obligation to provide shipment to our service center for repairs. We will provide return shipment by standard ground delivery. The customer may request second day or next day delivery at his/her own expense.

Returns

Return to the place where purchased.

It is recommended that any returns be made using original packaging or equivalent. When shipping be sure to **allow 2 inches of packing** on all sides for protection. **Always insure** for the full value of the package contents as packages get lost or damaged. Shipping cheaply can cost you your entire investment.

If returning any ParaZapper product to the factory, **call us before returning** so that we will be aware and watch for package arrival.

Broken wires

If wires are pulled from banana plugs simply unscrew colored barrel from plug, place wire through barrel and then through hole in the plug, wrap around the plug clockwise, then re-screw the barrel on to the plug tightly.
If the wire break inside of the copper paddles, call us.

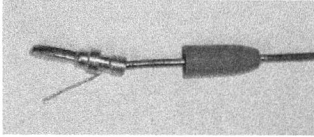

Wire feed through the end of the plug and out the side.

Repairs

Most repairs are free within the first five years and if charges are required, they are reasonable and economical.

Replacement parts

Call for any replacement parts or accessories that you may need.
We have replacement copper paddles available when needed.**If your zapper does not work as expected or if you have any questions, call, or e-mail us at** sales@paradevices.com**.**

Special Accessories

Zapping Pets

Generally, zapping your pet depends on the personality and attitude of the pet. Some individuals place their pets in their lap and hold the copper paddles against the animal's body, one paddle at each end. The fur should be wet enough to conduct the current without causing irritation.

We have available special pet paddles (3 inches by 2 inches) that can be placed on your dog or cat and held in place with an elastic bandage. They can also be placed under the animal when it is resting. The use of paper towels soaked in salt water is especially important so that the electrical signal passes through the hair to the body. Some users have trained their cats and dogs just to lay down on the paddles and rest.

Tropical fish

For aquaria without invertebrates, one copper paddle may be placed into the water at each end of the aquarium. Do not leave the copper in the water for more than 1 hour or so. **Note: Copper kills plants and many invertebrates.**

HORSES and Other Large animals

We have available special horse paddles (6 inches by 4 inches) that can be placed on your horse and held in place with an elastic bandage. These are often used on show horses, which must be antibiotic free. The may be held to shoulders and hips with elastic bandages or they may be placed under the hooves.

Special Brass Wrist Bands

We also have some special brass wrist and ankle bands (6 inches by 2 inches with sufficient strap to fit large wrists) for those who need the freedom that these allow. These are by special order and appear to work as well as copper paddles. They are definitely better than standard wrist straps. While there have not been any reported problems, it is possible that the metals alloyed in the brass may cause some problems.

Pet Pads   Brass wrist bands   Horse Pads

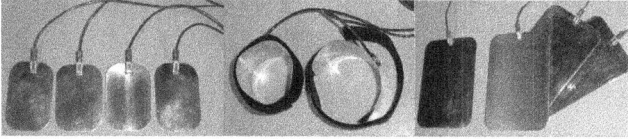

Using wrist straps

Wrist Straps used by many zappers are not as effective as other electrodes. These straps and cords while being convenient are made for anti-static applications and include a high resistance (1 megohm). This steals most of the voltage that the zapper puts out and makes it much less effective; especially those with coiled cords.

Note: The straps and cords sold for ParaZapper™ are low resistance (less than 1 ohm) to solve this problem.

Wrist straps are never as good as the copper paddles but may be used for maintenance after most parasites are gone.

For those who insist on using the wrist straps or for those who really need them, we offer **brass wristbands** which provide a better contact. When using either wrist straps or wristbands, be sure to wet your paper towel well with saline (avoid making it too strong or too weak) and wrap it completely around the wrist before putting the wristband on. The wrist strap has a square metal plate on the back, which should make full contact with the wet paper towel. Not following instructions as above may cause minor burning or rashes on the wrist.

IMPORTANT: Do not use the wrist straps with the footpads as wrist burns may result. This is due to the large difference in surface area, which can concentrate the signal at the wrist pad with the green lead.

56

Accessory Footpads

Footpads are available and help to generate a complete signal propagation throughout the entire body reaching areas that are more difficult to reach with the copper hand-holds or wrist straps. Users tell us that they improve results significantly, actually about double. Users can make their own footpads or even a single plate out of copper. Do not use aluminum or other metals as the may have serious effects. Use only C11000 alloy copper, which is pure.

You can also place a extra copper paddle under both feet in place of the footpads which will provide some benefit but not as much as the footpads.

When using the footpads and paddles, the green wire paddle will feel stronger because it is carrying 3 times as much current.

4 point contact - 3 points positive

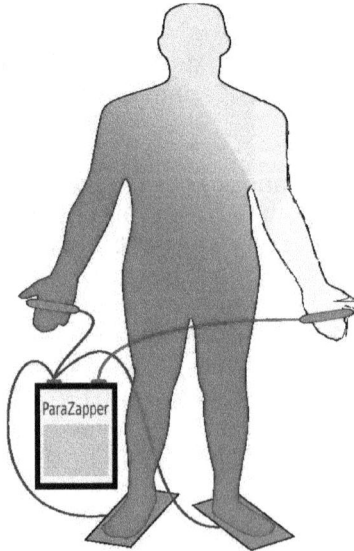

TENS Pads

TENS pads are readily available, convenient, and easy to use, however, they are also less effective because of the smaller contact area and higher resistance.

57

Batteries and Power

If the **Low Battery LED** turns RED then the battery is lower than needed for best results. This is between 8.0 and 7.5 volts. You can continue to use for a short period of time. Some competitors claim that batteries last longer by running them down lower. This may reduce the overall effectiveness of the zapper.

Batteries

There is no need to buy expensive batteries unless you are using the microprocessor based zappers such as the 6-Pack, CC2, UZI, or MY zappers. **Store brand heavy-duty 9-volt batteries** work just fine for the older models but the newer models need Alkaline batteries.

Do not use standard **rechargeable NiCad batteries** as they only output 7.2 volts at full charge. If you use rechargeable batteries, they must be rated for 8.4 to 9.6 volts at full charge.

The newer zappers need NiMH batteries rated at 8.4 volts minimum and 250 maH minimum.

Turn your ParaZapper off before changing batteries.

Status Indicator

The **standard ParaZapper** has a **single green LED** to let you know that it is on and does not have a status indicator. This zapper has a metallic label and does not have a model after ParaZapper.

ParaZapper PLUS and ParaZapper CCa only

If your ParaZapper has a status LED indicator, it serves multiple purposes. As the LED toggles from red to green as the output changes, the 2 colors combine to form a yellow or amber color when the zapper is operating properly.

You can also use this to check whether or not you are getting a good signal. Set the copper handles down, apart from each other and turn on the zapper observing the LED color. The color should be yellow or amber.

58

Now pick up the copper handles and you should notice a color change towards red. The more the color changes, the stronger the current that you are receiving. The color change is slight with the ParaZapper PLUS but is very noticeable with the CCa. Touching the paddles together will shift the color more toward red.

Foods, Nutrition, and Environment

One of the main reasons that some parasites become a problem in the human body is the loss of normal pH. The body normally has a slightly alkaline pH but most people who have serious illness such as cancer have an acidic pH. To help fight these illnesses, it becomes imperative that a normal pH be regained. The best way to do this is to have a diet that includes as much raw uncooked fresh vegetables and fruits as is possible. Additionally, these fresh raw uncooked fruits and vegetables provide many vitamins, minerals, and enzymes that we would not receive otherwise. The enzymes are especially important, as they are the essence of life.

This does not entirely preclude eating some meat. Two or three ounces of meat a day is sufficient to provide needed proteins. It is recommended that eating meat from animals that are not fed hormones and other chemicals is important.

There are also a number of nutritionists that recommend against dairy products. It is doubtful that 4 or 5 tablespoons a day of kefir or yogurt would cause significant harm. The absence of the needed bacteria can cause significant harm. See: **Intestinal Flora**.

Importance of Water

Our bodies are over 80 percent water and water is the medium by which nutrition is transported through our bodies and across cell membranes.

Water intake is also extremely important. While many bottled waters are better than many city water systems water, it is important to make sure that you are getting good clean water, free of any dangerous metals, microbes, and other potentially harmful contaminants.

Drink plenty of good clean pure water but do not over do it! **Rule of thumb:** Divide your weight in pounds by two. Drink this number of ounces of water a day.

Ozonated water is probably far better than chlorinated.

Vitamins

Vitamins are organic molecules that the body needs in order to function correctly and that it cannot manufacture on its own. The most efficient and reliable vitamins are in liquid form. When vitamin pills, capsules, etc are taken, the body only absorbs between 10 percent and 20 percent while 95 percent of the liquid vitamins are absorbed.

Vitamin pills often only contain a few isolated vitamins and as a result do not provide the complete nutritional support that is needed.

In reality, it is far better to get your vitamins and minerals from whole food concentrates as these contain other compounds that can assist in the utilization of vitamins. Fresh fruits and vegetables are the best source in many cases.

Exercise

Getting sufficient exercise is often mentioned but some important aspects are sometimes overlooked. Oxygen is critical to the support of good health and this is brought to the body by the circulatory system. If the circulation is sluggish and inefficient, then the body will suffer from insufficient oxygen transport ability. The same transport problems will also be apparent for vitamins, minerals, and other metabolic support. Any area of the body that does not receive sufficient circulatory support will suffer.

Sleep

Sleep is the state in which the body does most of its healing. It has long been known that people who do not sleep well often do not heal well. If you do not sleep sufficiently, get more exercise. If this does not help, melatonin often does help but do not use this as a substitute for exercise. Melatonin is the natural hormone that your brain produces to induce sleep but this production decreases with age and other causes. As a sleep aid, 1 to 3 mg just before bed is often helpful but some cancer patients take as much as 6 mg, two or three times a night with definite benefits. Up to 70 mg a night is safe in extreme cases.

Sunshine

Sunshine is the source of almost all life on the earth and it is critical for good health and well being. It has many benefits and a small amount daily is suggested. Sunlight produces vitamin D in the body, which is very important to good health. Being in the sunlight also provides a sense of well being that is also important to healing. Spend several minutes in both the early morning sunlight and late afternoon if possible

Fresh air

Recent studies have found that seriously ill patients who get exercise, fresh air, and sunshine have almost twice the survival expectancy as those who do not. They also have a higher chance of overcoming their illnesses.

Blue Sky

One additional benefit of being outdoors is the effect of blue sky. The blue color is noted to reduce depression and help provide stress relief.

Stress

Stress is a major cause of illness and some researchers have concluded that stress is responsible for as much as 80 percent of major illnesses. Stress can induce nutritional disorders as well as many common health problems. Find a way to eliminate or reduce the things that cause stress in your life.

Clean Environment

If your house is full of toxic materials and fumes, your health is not likely to improve. Even the very cleaning agents that you use may be causing your health problems. Carpets and walls leach toxic fumes that you may have become accustomed to. Scented candles, fragrances, glues, paints, polishes, etc can all contribute to unhealthy conditions and result in poor health. Clean your house thoroughly, removing anything that may have a negative effect.

Intestinal Flora

Zapping and the use of antibiotics can kill much of the normal bacteria that live in the intestines. If these are not replaced quickly and efficiently, unnatural pathogens and opportunistic microbes will take over. These unwanted bacteria, protozoa, yeast, and fungi can cause additional serious health problems if not addressed.

The best protection is to replace the lost intestinal flora with bacteria from kefir, yogurt, Smoothies, or other natural bacterial cultures.

Probiotic tablets

Probiotic tablets are not as good as the live cultured bacteria, as the tablet forms and even most liquid forms are in a suspended state and can take a couple of hours to recover.

Just make sure that the label of your kefir, yogurt, Smoothie, etc states that it contains live active cultures.

Kefir

The best probiotic is **kefir** as it contains 10 different types of beneficial microbes. Kefir may be found at Publix Super Markets in the US.

Smoothie

Second best probiotic source appears to be the **Smoothie**, which contains 6 different beneficial microbes. There is even a non-dairy form called a **Silk Smoothie**.

Yogurt

If you use **yogurt**, it should state that it contains three or more bacteria types in it.

Also good are kimshi (spicy Korean cultured vegetables much like sauerkraut), sauerkraut, and buttermilk.

Acidophillus

This is one of several beneficial bacteria that normally inhabits the intestines. Acidophillus alone is not good enough. You need at least 5 to 6 different bacteria strains, including Bifidus and Lactobaccillus.

Zapping and Metals

Users need to be aware that using various metals and having certain metals in their bodies can have a negative effect on their health whether zapping or not. Some of the more common of these metals are listed below.

Electricity and the transport of metals

Electricity is well known for its ability to move metal ions. This is the very principle of electroplating. Unfortunately, this process can cause serious health issues when certain toxic metals are placed on or into the body. A prosthesis made of stainless steel, as an example, can leach Nickel, Chromium, or other metal ions into the body under the influence of any electrical signal.

Aluminum

Aluminum has been associated with several neurological disorders and it is highly recommended that it not be used as an electrode for zapping. Avoid using aluminum cookware and aluminum containing products such as baking powder or aluminum silicate.

Brass

Brass is an alloy of Copper and Zinc. It can be alloyed with other metals for special properties. If this is used as an electrode, be sure to use CDA260 or other alloys with low contamination.

Chromium

Chromium is a serious tissue irritant and should not be worn, especially when zapping. Some bracelets and watchbands may be chrome plated. Chromium is also found in some stainless steel alloys.

Copper

Copper has been associated with cancer by Dr. Hulda Clark and others. The source appears to be mostly from copper pipe used for drinking water. The use of wet paper towels keeps most of the copper away from the skin when zapping. Use only C11000 alloy, which is basically pure copper (99.9 %). It is important to minimize copper contamination to your skin.

Gold

Gold is generally considered safe provided it is relatively pure and of good quality (18 ct). White Gold may have nickel or palladium in it. Green gold may have cadmium, which is toxic. 18k red gold is 75% gold and 25% copper and is the safest. Either 14k red or yellow gold is also acceptable.

Iron

Iron is basically a safe mineral that is easily handled by the body unless excessive doses are taken internally. Excessive iron can be fatal.

Lead

Lead is a very toxic metal, which can cause extensive physiological and mental problems. Because of this, ParaZapper uses lead free solder on its products. ParaZapper products are designed to be **RoHS** compliant.

Mercury

Mercury is highly toxic and is used in the common silver amalgam fillings used to repair tooth decay. Dr. Clark and many others strongly suggest that all amalgam fillings must be removed. While we have not seen any evidence that zapping will increase mercury in the body when amalgam is present, the user should pay careful attention to this. Some fish may have excessive mercury.

Nickel

Nickel is a serious tissue irritant and is often used as a plating material on watchbands and bracelets. These should not be worn when zapping. Nickel is also found in some stainless steel alloys.

Palladium

Palladium appears to be safe and innocuous for use in humans. Certain palladium compounds have demonstrated effectiveness in fighting cancers.

Platinum

Platinum is generally considered inert but if absorbed, small amounts can be toxic.

Zapping and Metals

Silver

Silver if it were not so expensive would probably be one of the best electrode materials as it has excellent antibiotic properties. An excessive amount of ionic Silver can be toxic though. Pure silver is often ingested as colloidal silver.

Stainless Steel

Stainless steel is an iron alloy that may have Cobalt, Nickel, and / or Chromium included in it and as these are tissue irritants, stainless steel should be avoided. Some users may have prosthesis of stainless steel and sometimes zapping can cause pain, swelling, and irritation around these.

Titanium

Titanium appears to be safe for use as an electrode and we have not had any reports of problems from zapping with implants or prosthesis made of titanium.

Hip and knee replacement

These prostheses are usually made of either stainless steel or of titanium. There have been several reported problems from those who have stainless steel implants but so far we do not have any reported problems with titanium. Some of the problems associated with stainless steel when zapping are sensitivity, itching, pain, and swelling. If you have a stainless knee joint but need to use the footpads, try placing the footpads on the thighs above the knee. The problem only seems

67

to occur when the electricity passes directly through the prosthesis.

Frequency Notes

Back in the late 1960's and early 1970's, singer Ella Fitzgerald astonished many TV viewers as she shattered wine glasses by singing a note with the right frequency for that glass. This is very much how the zapper works.

Another example is the case of the first Tacoma Narrows Bridge, which vibrated at 30 Hz when caught in a wind. This vibration destroyed the bridge in July 1940.

Complimentary Therapies

Many of the following therapies are combined with zapping to help produce additional benefits. Only a brief discussion is offered to introduce you to the possibilities.

Beck Protocol

The Beck protocol was developed by Bob Beck to help promote better health. It is composed of 4 main parts.
1) Blood electrification
2) Magnetic pulsing
3) Ionic or colloidal silver
4) Freshly ozonated water

Blood Electrification (BBBE)

This was originally developed by Bob Beck as a means of purifying blood and originally required that the blood must be removed from the body but the process was modified to treat the blood as it passed through arteries in the arms or legs. Electrification is one of the processes that are part of the Beck Protocol.

Cleanses and flushes

These are recommended by a large number of therapists and vary widely depending on the source. Dr. Hulda Clark recommends killing and removing parasites before flushing.

Chelation

Chelation is a process of binding systemic metals for the purpose of eliminating them from the body. Some of the products used are alpha lipoic acid, glutathione, Cilantro, EDTA, Bentonite clay, zeolite clay, and Chlorella.

Colloidal Gold

Colloidal gold is a solution of microscopic gold globules suspended in water and is reputed to be antimicrobial in effect.

Colloidal Silver

Colloidal silver is a solution of pure silver globules dissolved in water and is accepted as being a strong antimicrobial agent. Colloidal silver is different from ionized silver in that it is made from ion free distilled water and it definitely does have anti-bacterial and anti-viral properties. The issue of skin turning blue is mostly associated with ionic silver and taking very large amounts over an extended period of time.

Colonics

Colonics is a process of flushing out the colon with water or coffee and is similar to an enema but it is more extensive in reach and effects. It can be very beneficial to remove junk and toxins that have built up in the colon over many years.

Detoxification

Detoxification is the process of removing unwanted toxins from the body. There are many processes and products available for this.

Epsom Salts

Epsom salts is an inorganic chemical that serves multiple purposes including acting as a laxative and being used to clean out the intestinal tract. The advantage is that Epsom salts works as reverse flush which breaks material loose from the intestinal walls. Normal use is to place 1 tablespoon in a glass of water and drink. The taste is awful so it is helpful to chase it down with an ounce of lemon juice. You need to stay very close to a toilet for at least 8 hours even if it starts working immediately. Do not use Epsom Salts frequently. It is also good to bathe or soak in a tub with Epsom Salts. Do not take Epsom salts if there is any chance of bowel obstruction.

Essiac Herbal Tea

This tea reportedly has produced positive benefits to many people who have taken it to help with many illnesses. See: http://www.roadtohealing.com

Fasting

Fasting helps to clean out the intestines and to remove many products that have built up in the body over time. A water only fast of 24 to 48 hours is sufficient but many fast even longer. One of the fastest ways to shrink a tumor is fasting.

Frequency Generator

The frequency generator is a device that can produce specific selected frequencies. It is often used to target specific parasites in the body rather than using general frequencies like the zapper does.

Hydrogen peroxide

The addition of small amounts of food grade hydrogen peroxide to drinking water has many benefits, including killing unwanted bacteria and removing harmful contaminants.

Iodine

Iodine has been shown to be extremely important to the function of the human body and many humans are at least moderately deficient.

Ionic Silver

Ionic silver is a solution of silver ions that are combined with other elements. If silver is dissolved in water with chlorine in it, it will form silver chloride.

Juicing

Crushing, liquefying, and extracting the juices from vegetable has been shown to have extreme health benefits. Most importantly, juicing provides live enzymes that the body depends on.

Kidney Cleanse

Dr. Clark recommends using an herbal tea for flushing out the kidneys and cleansing them. The more efficiently your kidney function is, the better off the rest of your body will be.

Kombucha tea

This is a fermented black tea reputed to have beneficial properties. As it can leach metals, keep it stored in glass containers for best results.

Live Food

Live foods contain active enzymes that the body needs to function correctly. Because we often eat so many cooked and prepared foods, we do not get the supply of active enzymes that our ancestors did.

Liver Cleanse

Dr. Clark's Liver cleanse is one of the most popular of many different cleanses and flushes for the liver. It helps remove gallstones and other contaminates from the liver and gall bladder. It is strongly recommended to remove parasites first as they can interfere with the cleansing action. An advantage of liver cleansing is improved handling of cholesterol by the body.

Magnetic Pulser

The magnetic pulser is another part of the Bob Beck protocol and is used to reach parasites that may be hiding in the lymph system and internal organs.

Noni Juice is a NO NO!

The main effect of this product appears to be an energy rush that is a result of the high levels of sugars that it contains. It should not be used with diabetics. The excessive sugars are also bad for Cancer, Candida and other problems. There are some Noni products that do not contain a lot of added sugars, etc. These may be beneficial, but do not taste or smell good.

Ozonated water

Ozone is recognized as an effective microbial agent and additionally is reputed to increase the effectiveness of oxygen transport in the blood stream. Various machines are available for adding ozone to drinking water for health benefits.

Ozonated oils

Ozonated oils, especially ozonated olive oil are considered to be very beneficial.

Parasite cleanse

The parasite cleanse is usually accomplished by ingesting several mildly toxic herbals that kill parasites in the body. Some of the herbals frequently used are wormwood and green tincture of black walnut. Using these cleanses combined with zappings should produce better results than zapping alone. We recommend Dr. Clark's parasite cleanse formulas.

pH Balance

The pH balance of your body is very important to good health and healing. There are several types of pH test strips available and you should use these frequently to make sure that you are within the normal range. Urine pH and saliva pH should be tested in the morning when rising and again in the evening. It may also be tested before and after meals to produce a better idea of your body's pH buffering capacity. We recommend pH test strips that have a dual color indicator, as these tend to be more reliable indicators. There are numerous sites on the Internet for pH testing supplies.

Plate Zapping

This method is preferred by some and uses samples of various tissues and organisms to concentrate the energy to affect those specific organs and organisms.

Raw Foods

Raw foods are known for having active enzymes that we need. The trade off is a chance for contamination.

Rife Therapy

Raymond Rife was among the earliest to explore the relationship between sonic frequencies and microorganisms and he found that almost any microbe could be killed by a specific frequency. His early work used analog radio equipment and transmitted the frequencies through the air.

Synchrometer

The synchrometer was introduced by Dr. Hulda Clark and is used to detect the various types of parasites living in the body. The device is very sensitive and requires a lot of experience in order to produce reliable results.

Terminator

The Terminator used to be the most popular zapper on the Internet because of its convenience and reputation. One advantage that it had was lower frequency, which had stronger effects on some parasites than the Hulda Clark zapper did. The disadvantages were that the electrodes were small and close together.

Urine Therapy

Urine therapy is growing in popularity and is based on the idea that not everything passed out in the urine needs to be discarded. An example is the case where antibodies may be discarded in the urine but if the urine is recycled, the antibodies are re-absorbed and reused. Urea has been shown in some cases to have definite health benefits.

Water Cure

The basis of the water cure is that many people do not drink as much water as they should. The general principle is that a person should drink half of their weight in ounces of water daily. A person who weighs 150 pounds should drink 75 ounces of water each day. Also, it is generally accepted that ¼ tsp. of sea salt should be added to each gallon of water.

Wheatgrass Juice

Drinking the juice of wheat, grass, or barley provides a lot of antioxidants including chlorophyll as well as trace minerals. It is considered very beneficial for cancer and other illness.

Reactions

About Die-off

Die off happens whenever foreign organisms die inside of your body. It does not matter what causes the organisms to die and it is not unique to zappers and zapping. The same effect results if chemicals, drugs, or herbals are taken. A die-off reaction is definite evidence that you are getting results.

Herxheimer reaction

Often referred to as a Herx or Herxing, it is a reaction caused by microbes dying and releasing toxins faster than the body can handle it. Basically, herxheimer is a strong die-off reaction creating a sudden and exaggerated inflammatory response. If the results are too strong, you might stop for a day until you adjust. Some choose to zap again to improve results.

Rashes, itching, stinging, tingling sensations

These are normal symptoms of die-off and can be expected in many cases but not all. It is wise not to panic but the user should be aware that these may be symptoms of other problems and if they persist, a medical professional should be consulted.

Electrodes

Oxidation of electrodes

It is important to keep your electrodes clean and shiny as even a slight brown or gray oxidative coat can increase the contact resistance and reduce the effectiveness of your zapper. For best results, polish your electrodes (copper paddles and footpads) daily before each use.

Avoid the use of Stainless Steel which contains Chromium, Nickel, or Cobalt.

History of Electric Healing

This is presented as a basic history and is in no way complete. That would be quite a large book in itself.

The use of electricity goes nearly back to the first re-discoveries of the nature of electricity itself in the mid-1700's.

Although electricity was somewhat known and used 4000 to 5000 years ago, there is little mention of it from early history up to its re-discovery in the 1600's.

The **ancient Egyptians** may have utilized low voltage devices as far back as 2000 B.C. There are Hieroglyphs that show the ancient Egyptians holding what were long described as scrolls of papyrus but these may have actually been metal cylinders as remnants of these have been found in tombs. Holding these dissimilar metal cylinders in the hands could turn the body into a battery, producing a slight voltage differential across the body. These have been referred to as the Rods of Ra or the Horns of Horus.

Around 600 BC, the Ancient Greeks discovered that rubbing fur on amber (fossilized tree resin) could cause an attraction between the two – and so what the Greeks discovered was actually static electricity.

Baghdad Battery

Remnants have been found of ancient electric batteries known as the **Baghdad Battery** that originated sometime between 250 B.C. and 225 A.D. but some other believe that they may have existed long before that. Some believe that these were used for electro-plating but others feel that they may have been tried for healing purposes.

Romans and ancient Greeks had used animal electricity (electric ray) for medical treatment during Biblical times. As far back as 63 A.D., doctors were able to treat migraines and even epilepsy using electro-therapy.

The modern history of medical electricity begins with the inventions of the first practical static generator in 1742 and the Leyden jar for storing electricity in 1745. Doctors of the day electrified patients with static electricity or gave them strong shocks. This produced some promising results, but the technology was incomplete.

William Gilbert became the chief physician to Queen Elizabeth and is credited as one of the originators of the term "electricity" and wrote about it in his book **De Magnete** (approx. 1600). Thomas Browne, wrote several books and he used the word "electricity" to describe his investigations based on Gilbert's work.

While **Benjamin Franklin** is often referred to as the Father of Electricity, his "kite-lightning experiment" did not take place until 1752. By then, the use of electricity as a healing aid had already been well established in England.

Richard Lovett, after years of practice, published the first English textbook on medical electricity in 1756, which soon became essential reading for any electrical healer.

Shortly following, **John Reddall**, who asked Richard Lovett for advice for planning a course of lectures on medical electricity. Reddall's London lectures were so popular that they attracted more than a hundred people a day.

At that time, medical help was mostly available to the wealthy and to the privileged. For that reason, the practitioners of electrical healing worked hard to make electrical healing so cheap as to be affordable to the poor and lower classes.

This was one of the main concerns of **John Read**, who had been a patient of John Wesley. He realized the importance of making the apparatus portable and less expensive, so he trained himself by attending some of Lovett's demonstrations. Read's electrical machine soon became a standard instrument and was acknowledged by Joseph Priestley as practical for medical purposes.

In 1760, **John Wesley** published "The Desideratum, or Electricity Made Plain and Useful". Previously, In 1747 he published "Primitive Physick, or an Easy and Natural Way of Curing Most Diseases", which was a very successful pamphlet addressed to the lay public that contained a list of over nine hundred recipes for medical remedies and practical directions on how to cure a large number of disorders.

Also, in 1746, Wesley opened several dispensaries both in London and in Bristol to provide free and inexpensive services to the poor and needy. Ten years later as he developed interest in electric healing, he also began to offer free electrical treatments.

During this period, there was also great philosophical struggle concerning the use of electricity. The electrician **Tiberius Cavallo** lamented that electrical therapy was not as effective as it might have been due to insufficient knowledge and understanding of the medical practitioners who employed it. Paola Bertucci has shown in a stimulating thesis on medical electricity, Cavallo obtained much of his medical knowledge from close medical friends such as the physician James Lind and the surgeon Miles Partington.

In 1794, **William Hawes** founded the London Electrical Dispensary, a charity especially conceived to offer free electrical treatment to the poor.

John Fell, invested a considerable amount of money in textbooks and instruments for performing electro-therapy treatments, cured forty three patients.

The interest in electric healing continued to grow.

n 1780, **Luigi Galvani**, an Italian surgeon, observes the effect of electrical current on animal tissue and with subsequent experimentation produces a paper in 1791 entitled "The Effects of Artificial Electricity on Muscular Motion".

Batteries began to be available in the early 1800s which revolutionized and accelerated the use of electricity for healing.

In 1831, **Michael Faraday** discovered that magnetism can be used to produce electricity. This lead to development of better electric production and was also the basis for the discovery of radio waves, X-rays and the electron microscope. Michael Faraday's experiments also opened the door to alternating current. AC current was only a curiosity at first, and then it was adopted for the power transmission grid, mostly due to the work of Nicola Tesla.

In approx. 1836, **Guy's Hospital of London** set up an "electrifying room" where patients often sat on an insulated stool and received an "electric bath" from a "static machine." Doctors drew sparks from the electrified patients or shocked them with Leyden jars.

Edward Hartshorne graduated from the University of Pennsylvania School of Medicine, Philadelphia in 1840. While working at Insane Department of the Pennsylvania Hospital, he described the use of electrical stimulation by John Birch, in 1841 to treat a tibial non-union in a patient who underwent treatment in 1812 for a tibial nonunion with 'shocks of electric fluid passed daily through space between the ends of the bones.'.

Rev. Nicholas Callan invented the induction coil which later led to the transformer. Published "On A New Galvanic Battery" 23 August 1836. This laid the ground for Faradic devices and the Medical Electric Battery which was developed later.

Subsequent work by **R.W. Lente** in 1850 further recognized the potential for electricity to heal bone, when he presented a series of case reports describing electricity's effect on fracture non-unions and pseudoarthroses.

In 1853, the British journal, Medical Times and Gazette, published an article, "Galvanism to the Ununited Fracture", that described the use of electrically charged needles inserted into a fracture site to cause healing.

In 1872, **Dr. A.D. Rockwell**, who later contributed to the development of the electric chair, requested to read a paper on recent research in therapeutics, before the New York Medical Society, but was turned down because electricity was considered to be the domain of crooks. By 1890, five medical schools in New York were teaching courses in electricity. There was a great wave of interest in using electricity for medical treatment, which lasted until the 1930's.

It was about this time, in the 1870's that galvanic and Faradic devices started to become popular. The Electro Magneto Machine was patented by **Thomas Hall** in 1859 and was one of the earliest machines of its type, although there were many different versions manufactured by a number of companies.

McIntosh Galvanic and Faradic Battery, established 1879, pictures show a large building with a large number of production workers. They manufactured a wide range of Faradic machines and even had electrodes for "Tub Zapping", as well for almost every part of the human anatomy.

Dr. Guillaume Duchenne (de Boulogne) moved to Paris in 1842 where he practiced 'electrisation' for a number of years, keeping extensive notes and writing a number of publications. A book on his works is published in Google Books.

High-frequency alternating currents arrived in the 1890s with the suggestion that they would be valuable in medicine.

Nicolas Tesla

In 1892, **Nicolas Tesla** presented lectures across Europe and met with **Paul Oudin** in Paris where they discussed ways of building electro-therapeutic devices. Paul Oudin built the first "violet ray" and wrote an article on using it to cure skin disorders the next year.

Tesla published a paper in 1898 that he read at the eighth annual meeting of the American Electro-Therapeutic Association, entitled, "High Frequency Oscillators for Electro-Therapeutic and Other Purposes."

The name "violet ray" occurs for the first time in 1913 in a dental journal. Paul Oudin began to experiment
81

with skin disorders and found that acne, eczema and psoriasis were easily treated with the new device. After a few treatments the skin patches would begin to break up and disappear completely in two to three months. When the devices were used to spark warts or skin cancer, the anomalies often were removed within weeks.

The violet ray often took away pain, and many times it was almost considered a miracle. It was valuable in dealing with arthritis and was often considered a miracle in rheumatoid arthritis.

The era of electrical healing lasted from the early 1890s to some time after 1910. A conspiracy to limit and eliminate competition from non-drug therapies began with the Flexner Report of 1910 when Abraham Flexner was engaged by **John D. Rockefeller** to evaluate the effectiveness of therapies taught in medical schools. Anything that used electricity was quickly labeled as quack practice.

However, by 1916, inexpensive violet ray units were being sold in drugstores under this name, along with products such as the **Voltamp Medical Electric Battery**. The Medical Electric Battery was basically, an early form of zapper. A dozen companies manufactured the violet ray tubes in the United States and several other countries. Drugstores had front window displays of violet rays. Hundreds of thousands of violet rays were sold and used, with few reported problems. These products were extremely popular from the 1890's to the 1930's, being found in many catalogs such as Montgomery Wards and the Sears Roebuck Company.

When they became popular with the public, doctors and the FDA started to despise them. At first the Journal of the American Medical Association published promising therapeutic results in articles. Then it printed an article about a man who deliberately short-circuited his violet ray and electrocuted himself. This implied that the device was dangerous and should be outlawed. This is actually typical of the actions taken when the medical industry wants a competing product removed. They did the same thing with a man who used a Hulda Clark zapper when his pacemaker failed. It was not a flaw of the zapper but was instead poor design of the pacemaker. They also did a similar action with a person in Australia who was actually improving by using a zapper but stopped and later died. They called it a Rife machine, which it was not in any way. This is the normal pattern of denigration used against products that the AMA, FDA, and much of the medical industry do not want people to find.

The violet ray in healing would have been almost totally forgotten, except for one man. Around 1900, Edgar Cayce, at the age of 24, lost his voice for months and doctors were unable to help him. His recovery under hypnosis is a tribute to the power of suggestion. While he did not use it for his healing in this case, he did recommend the use of the violet ray in over 800 readings.

The violet ray and the Faradic Battery are both grandfathered devices, meaning that it was produced before 1976. According to Congressional law, grandfathered devices are generally presumed to be safe and not subject to federal regulation. In spite of this, the FDA ignores this and has threatened legal action against the companies that produce them and the people who use them.

On Sept. 13 to Sept. 15, 1898, Tesla also gave a lecture titled "High Frequency Oscillators for Electro-Therapeutic and Other Purposes." where he introduced the MWO or multi wave oscillator.

Georges Lakhovsky

Georges Lakhovsky was a Russian who lived in Paris in the 1920's who compiled his observations about the effect of electricity and radio waves on living organisms in "Curing Cancer with Ultra Radio Frequencies" in Radio News, which he first published in 1925. **Lakhovsky** also built a device he called a Multiple Wave Oscillator that was based on Tesla's unit. The purpose of this device, a very short-range high voltage transmitter that broadcast a multiplicity of frequencies at once, was to induce currents and voltages of very high frequency into living organisms for the treatment of disease. At least, that's what it

83

was supposed to do. Lakhovsky's first units simply didn't work.

Frustrated by the issues, Lakhovsky asked Nikola Tesla for help. Tesla went to Paris that year and reworked the circuitry in Lakhovsky's oscillator. Basically turning it into a Tesla Multi-Wave Oscillator. Tesla believed that, his high-potential, high-frequency currents could be passed into the body harmlessly, "these currents might lend themselves to electro-therapeutic uses." Tesla as a result of being struck by a Taxi, experimented upon himself, went to his hotel room where, with the help of his own electro-therapy, he recovered from his injuries. Tesla never patented any electro-therapy devices but did publish his observations in technical journals, and several years later he gave a speech to the American Electro-Therapeutic Association in which he details with drawings the high-frequency apparatus he invented for this purpose, which included a Tesla coil. Tesla's suggestions were taken up in earnest by George Lakhovsky, who perceived that the twisted-filament, coil-like structures within all living cells constitute ultra-microscopic circuits "capable of oscillating electrically over a wide scale of very short wavelengths."Lakhovsky's apparatus evolved from Tesla's. "These circuits," Lakhovsky wrote, "are stimulated by damped high-frequency currents from a spark gap. Thus each circuit of the transmitter vibrates not only on its natural frequency, but also on numerous MWO harmonics." The frequencies of his oscillator's basic vibrations **ranged from 750 kilocycles all the way up to 3 gigacycles!**

Lakhovsky obtained remarkable results from a seven-week clinical trial performed at a major New York City hospital and that of a prominent Brooklyn urologist in the summer of 1941. Later editions of The Secret of Life detailed many of these cases.

It was reported that Georges Lakhovsky had a 98% success rate in treating fatal cancers over an 11-year period.

Electro-therapy was a standard treatment modality at the beginning of the twentieth century, but interest was lost in the years to come, mostly due to suppression by the AMA and the FDA. The violet wave disappeared off of store shelves along with the medical electric battery and they also disappeared out of the catalogs by about 1936.

In 1941, the Voltamp Battery #7 was found to be "misbranded" according to the FDA and the company was shut down by court order. This was basically, the end of the older electro-medical devices. Misbranded is basically the FDA's way of shutting down a claim of effectiveness when it has not been proven to their satisfaction.

For those who are interested in further research, a great collection of electrical healing devices can be found in the Bakken Museum of Minneapolis. Also, the Museum of Questionable Medical Devices in Minneapolis has some of the same devices. The Indiana Medical History Museum has a wide variety of electrical gadgetry used to cure disease.

Morris Fishbein

In 1924, **Morris Fishbein** became the head of both the AMA and the JAMA. His rose to power by labeling natural healers, Native American healers, and American midwives of the time as all being "quacks," empowering the chemical medicine industry that was to become the allopathic medicine that we know today. Doctors were reported to be coerced into joining the AMA for fear of losing their license. During his reign, Doctors who used or suggested natural cures were removed from their practices, had their licenses removed, their medical files destroyed, and some were never heard from again. Many encouraging cancer treatments were buried or denigrated. He also made sure nobody knew who Royal Rife was, and that nobody could buy the Rife frequency machine, a holistic treatment for cancer and infectious diseases. Even today, the FDA seeks to destroy the remaining original Rife machines. Also, he tried to bury the **Hoxsey** cancer cures, treatments that cured several forms of cancer, that had no destructive side effects. State medical boards regularly closed free clinics where terminal cancer patients were being saved by natural remedies.

Today, prescription drugs alone are responsible for at least 100,000 deaths of Americans each year, and the AMA seal still carries with it the same weight and significance.

Morris Fishbein was eventually convicted of racketeering charges. It is my opinion that he should have been charged with crimes against humanity because his actions eventually caused more deaths than the killing fields of Cambodia and probably mode than all the deaths in World War II. There are so many untold deaths and uncounted deaths that have occurred because of the denial of knowledge and the right to be treated according to ones own needs.

Rife

In 1934, the University of Southern California appointed a Special Medical Research Committee to study 16 terminal cancer patients from Pasadena County Hospital that would be treated with mitogenic impulse-wave technology, developed by Royal Raymond Rife. After four months the **Medical Research Committee reported that all 16 of Rife's formerly-terminal patients appeared cured**.

Dr. Rife was using his machines back in the early 1920's as the following shows.

"In January 1920 experiments were started at the Rife Research Laboratory by Commander Royal R. Rife U.S.N. Ret. to determine the effect of electrical influences upon pathogenic microorganisms. Tests were made for anode and cathode polarity influences and the effect of infrared, ultraviolet and X-ray. During these experiments the idea was conceived of the possibility of devitalizing the pathogenic micro-organisms by electrical frequencies of varying wavelengths. The initial apparatus (Rife Ray #1) for the tests along this line of experiments was constructed and used in prolonged experiments during 1921 and 1922, with results that warranted the belief that the principles involved contained possibilities."

86

From:
http://rifevideos.com/chapter_6_dr_rifes_1920_to_1922_
rife_ray_1_rife_machine.html

There were numerous test performed to determine the MOR (mortal oscillatory rate or frequency) of many microbes. It was found that after blasting those microbes with the right frequency, the microbes would become lifeless and could not reproduce.

These experiments were reproduced extensively with consistent results.

For full information, see http://rifevideos.com

There are also a number of associates who worked with Royal Rife over the years, including Dr. Milbank Johnson, M.D., John Crane, John Marsh, Dr. James B. Couche, M.D., Dr. Lara, M.D., Dr. Robert P. Stafford, M.D., and others.

Dr. Robert O. Becker, M.D.

Dr. Becker performed some of the most serious scientific studies on the causes and effects of electrical function in biological organisms. After publishing a number of studies in various journals, he wrote and published two books on the subject and his studies. The first is "Body Electric" and the Second was "Cross Currents". Both are highly recommended reading to anyone with an interest in electro-biology or electro-medicine.

Hulda Clark

Hulda Clark published a series of books starting in 1992 and going forward to about 2006. Her first book, The Cure For All Disease" gave her notoriety and all of her books that followed were quite popular.

The titles were bold and gained a lot of attention and drew the wrath of the medical industry. The odd thing is that there were many criticisms placed by people who never bother to read the books.

At the very minimum, her books gave hope to many with terminal illness and in actuality, many had exceptional recoveries. There were some cases that did not have the best results and there may have been some significant reasons for it.

Bob Beck

In the 1990's Dr. Bob Beck developed a protocol that included a zapper that was different from Dr. Hulda Clark. The Beck Zapper produced a stronger voltage output than the Clark zapper and had a lower frequency of only about 4.0 Hz. It was actually supposed to be 3.9 Hz which was half of the Schumann Earth Resonance. However, the Beck Zapper or Bob Beck Blood Electrifier (BBBE) was not fully effective on its own as it was intended only to zap the Blood. The Beck protocol required a magnetic pulser to stimulate the Lymph system, and Colloidal Silver as an anti-biotic. There is much about Bob Beck on the Internet.

Scenar

Nicknamed the 'Star Trek' device, the *SCENAR* device was developed as part of a secret Soviet space program with the goal of maintaining the health of *astronauts* in space. The original *SCENAR* device was invented and designed by A. Revenko and A. Karasjev and was produced by ZAO OKB RITM.

How the SCENAR works

The SCENAR uses biofeedback, enabling the body to heal itself. The SCENAR sends out a series of signals through the skin and measures the response. Each signal is only sent out when a change is recorded in the electrical properties of the skin, in response to the previous signal.

Results form the SCENAR

In Russia, some 600 practitioners currently use the SCENAR as their principal treatment instrument, with over 50,000 reported cases of individual use. A vast wealth of information on the SCENAR is available from research papers, clinical reports and training manuals. The SCENAR can be used on most types of disease or injury: circulatory, sensory, respiratory, neurological, genito-urinary, musculoskeletal, gastro-intestinal, endocrine, immune and psychological disorders. The SCENAR is also credited with vastly reducing recovery times.

What has been missed in history?

The one common thing that has been overlooked throughout the entire history of using electricity to fight disease and illness, if the amazing ability of many of the devices to quickly destroy microbes such as bacteria, protozoa, fungi, and other microbes, possibly many if not all forms of virus.

Has there been any studies to support this? Quite possibly but you can be assured that any such reports have been excluded from publication in any medical journals as the journals are controlled almost exclusively by the pharmaceutical industry.

The evidence is quite obvious and compelling as there are videos and photographs that are easily available. Anyone who has a microscope can easily set up a slide of living microbes and observe for themselves the effects that electricity of various forms will have on differing microbes.

When observed under the microscope, it can be seen that the effect depends on frequency, voltage, current, and waveform applied. When the right frequency of sufficient voltage and current and waveform is applied, some microbes are completely destroyed in seconds while others may take several minutes.

Not only can microbes be killed but it is quite possible that other effects may be involved. For one, One seller claimed that his zapper would not kill

89

microbes, because it used a weaker output, but that it would still significantly boost the immune system.

This effect, while not established by research is supported by the claims of many and it may be the result of the immune system sensing the vibrational frequency starts seeking possible causes and attacking microbes that were previously ignored.

Pz00020,21,22   Standard ParaZapper

We do not manufacture these any longer but include the information because some competitors make similar units.

The switches are located on the end of the case where the cables plug in and are black push buttons. The **power switch** is on the left side of case with the label facing up. **To turn it on, press it in**. Pressing it a second time will release the switch to pop back out turning the unit off.

The **frequency selector** is on the right side of the case. For 2.5 kHz release it out toward the front edge. For 30.0 kHz press the switch in.
The green LED (light) will come on when the unit is on.

Note: Older units had a Red LED. These units have the standard output of the Original Hulda Clark zapper and should be updated.

pz00021 ParaZapper with wrist straps and copper paddles.

Pz00030,31,32 ParaZapper PLUS

The switches are located on the end of the case where the cables plug in and are black push buttons.
The **power switch** is on the left side of case with the label facing up. In order to **turn it on, press it in**. Pressing it a second time will release the switch to pop back out turning the unit off.

The **frequency selector** is on the right side of the case. To use 2.5 kHz, release it out toward the front edge, and for 30.0 kHz press the switch in.

The **multicolor LED** (light) will come on when the unit is on. When the unit is working correctly, it will show both red and green at the same time and appear yellow, amber, or orange. All Green or all Red condition indicates a failure. **See: Status Indicator**.

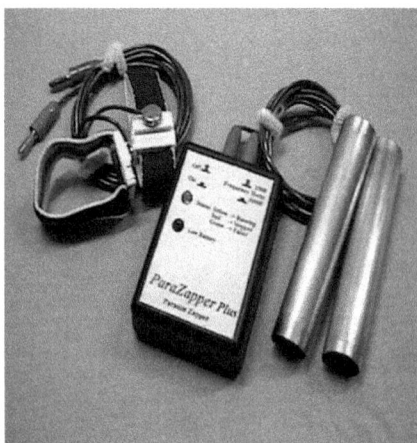

Pz00040,41,42 ParaZapper Cca

The controls on **ParaZapper™CCa** are on the front and adjust the current or signal intensity. The best results will be obtained by using the maximum level that can be tolerated.

The **switches** are located on the end of the case where the cables plug in and are black push buttons. The **power switch** is on the left side of case with the label facing up. In order t**o turn the zapper on, press the power switch in**. Pressing it a second time will release the switch to pop back out turning the unit off.

The **frequency selector** is on the right side of the case. To use 2.5 kHz, release it out toward the front edge, and for 30.0 kHz press the switch in.

The **multicolor LED** (light) will come on when the unit is on. When the unit is working correctly, it will show both red and green at the same time and appear yellow, amber, or orange. All Green or all Red condition indicates a failure. **See: Status Indicator**.

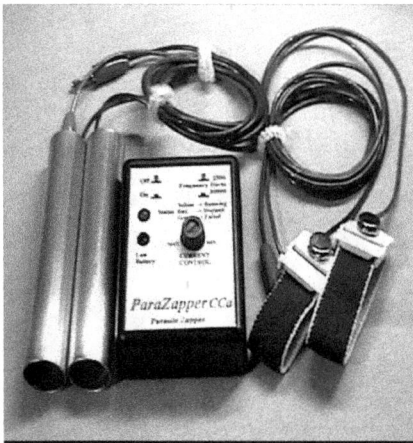

Pz00050,51,52 ParaZapper MX

While this zapper is not manufactured any longer, it is included as a basic mode in several of our models.

The ParaZapper™ MX model is programmed to output a sequence of 8 frequencies near 30,000 Hz and 8 frequencies near 2500 Hz. This unit has been optimized to assist with the reduction of HIV, AIDS and other viral infections as well as parasites. The zapper is connected and then turned on until it beeps (approx. every 14 minutes) for each zapping. A resting period of 16 – 20 minutes should be allowed between each zapping with a minimum of 3 zappings per session. If the ParaZapper™MX left on in the continuous mode, it will repeat the 14-minute cycle and beep at the end of each cycle. Every other cycle produces a different set of frequencies with the second cycle producing 1 beep.

Control knob on ParaZapper™MX models

The controls on ParaZapper™MX models are on the front and adjust the current or signal intensity. Using the maximum level that can be tolerated will produce maximum results. **Turn Clockwise to increase.**

The **power switch** is on the left side of case with the label facing up. In order to **turn the unit on, press the switch in**. Pressing it a second time will release the switch to pop back out turning the unit off.

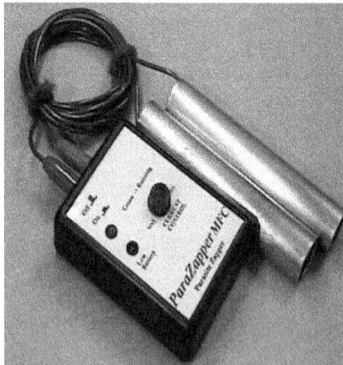

When first starting with the MX, it is recommended to zap one session a day for three or four 14-minute cycles at 30 kHz for the first 3-4 days. The destruction of parasites can place a heavy load on the body. After this initial period of 14 minute zappings, the user may increase their sessions to 2 or 3 28-minute zappings each day. Following this initial session, the user may zap additional zappings of 14 to 28 minutes once or twice a day. Do not forget to rest 15 to 20 minutes between zappings and it is recommended to always zap a minimum of 3 times for each session.

There may be an advantage to adding several short zappings (7 minutes minimum) a day once the initial phase is concluded. When using these short follow-up sessions, it is probably not necessary to zap more than 1 zapping each time provided that at least 1 full session has been completed that day and no more than4 hours has lapsed between zappings.

Those involved in clinical programs are requested to keep records of zappings in the back of this book or some other form. Include any background information possible and any test results. Also any personal observations and difficulties should be recorded.

Any serious complications should be reported to both the project team and to Para Systems and Devices as soon as possible.

When first turned on the MX will beep and both red and green LEDs should be on. After this initial period, the red LED should turn off. The green LED (light) should flash on and off when the unit is on and the unit is working correctly. The Red battery LED indicates a low battery, and continuous beeping may also be a low battery.

Because of the computer chip in the MX, power usage is higher and batteries do not last as long as some other units.

NOTE: ParaZapper™ MX2 is much stronger than other zappers and the user can expect to feel a definite strong signal.

Using the ParaZapper™AV

This model is not made any longer.

It is important that anyone who plans to use ParaZapper AV should use one of the standard ParaZapper products for two weeks continuous before starting to use ParaZapper™AV. The reason for this is that you need to eliminate regular parasites first before trying to tackle your viral illness. The ParaZapper™AV is not designed for the elimination of general parasites.

Starting with ParaZapper™AV

After zapping for a minimum of 2 weeks, preferably with ParaZapper™CCa, the user may start with ParaZapper™AV. The user should gradually ease into using this unit as the results may be harsh at times.

On the first day, use the unit for 2 hours. Place paper towels around the wrists and ankles then place the wrist bands over these and strap them on with the provided velcro. The wires may be worn under the clothing for discretion and the electronic unit is worn under the belt or waist band. Turn the unit on and allow it to continue for 2 hours 20 minutes. It will automatically cycle through the preset frequencies. After 2 hours 20 minutes, turn it off. The user can expect a die-off reaction that is more extreme than normal zapping.

On the second day, the user should repeat the same actions as the first but allow the zapper to run for 5 hours.

On the third day, the zapper should be allowed to run for 7-1/2 hours.
On the 4th day, the zapper should be allowed to run for 12 hours.
Replace the battery. On the 5th day and thereafter, the unit should be on for 24 hours a day except for when showering or during other activities that might cause damage to the zapper.

At any time that any adverse reactions are noted, a comment should be placed in the notes section of the manual and as soon as possible the factory should be called and notified. A toll free number is on the back of this manual.

Warning ParaZapper™AV **may cause painful burn marks on the wrist and ankles due to its signal strength and extensive usage.**

The battery should be replaced daily at the same time for maximum benefit.
There may be problems associated with the salt water and brass bands when used for an extensive period. This includes rashes and possible skin sores and is due to continued dampness.

Pz00076   ParaZapper AV

Rife Frequency Mode zapper for Virus

This model is not made any longer.

As this device is entirely experimental, the user will be using it at entirely their own risk. It is a new device and the short term and long term effects of use are not known. The user also agrees to notify the manufacturer of any difficulties or change of condition during the use of this device.

The Rife frequencies are of additional benefit against certain infections such as HIV, HEP-C, HEP-B, HHV, Epstein-Barr, as well as other viri. It is designed to be worn around the clock by those who are experiencing degenerative conditions. It is not a cure-all and its recommended use is only for those who are out of other medical options.

Timer

The **ParaZapper™AV** uses a timer routine that provides measured supply of frequencies that are reported to have the greatest effect on viri on an interrupted basis around the clock. The intention of this is to constantly kill new viri as they are released and before they can invade new host cells. It is felt that the user should obtain a noticeable reduction in viral counts over a 30-day to 60-day period. Results are not guaranteed, but if you feel that the unit is not beneficial, call us.

Wrist and ankle Bands

The included wrist and ankle bands are included to provide a good safe electrode with sufficient contact area to produce beneficial results. One band should be worn on each ankle with a red wire attached and one with a red wire on one wrist and a green wire to the other wrist. The red and green wires on the wrist should be alternated daily or more often. It is preferred to wear a damp paper towel under the wristbands to avoid metal contact with the skin. Using salt water to wet the paper towels will provide better
100

conductivity and should increase results. Covering with plastic should help retain moisture.

Frequencies

The unit produces a total of 10 frequencies and 2 rest periods over a 2 hour plus 40 minute period. The minimum effective period of use is 2 hours 20 minutes.

Pz00076   ParaZapper AV

Control knob on ParaZapper™AV models

This model is not made any longer.

The controls on **ParaZapper™AV** models are on the front and adjust the current or signal intensity. Using the maximum level that can be tolerated will produce maximum results. **In order to increase the strength, Turn the knob Clockwise.** If irritation or discomfort is present, turn the control down.

NOTE: Users should expect to feel occasional jumps in intensity as it cycles through each frequency.
The **power switch** is on the left side of case with the label facing up. **Press the switch in to turn the zapper on.** Pressing it a second time will release the switch to pop back out turning the unit off.

A green LED (light) will flash when the unit is on and the unit is working correctly. A flashing Red battery LED indicates a low battery.

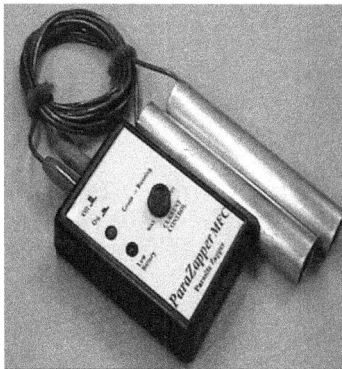

Other Zappers

The $10 Zapper from zapperplans.com

This zapper is a cheaply manufactured zapper for those who can not afford anything else. It does work for some problems but from my experience, it pales in comparison to other zappers.

The first problem is that the chip used in it (CD4069) is not as good for zapping applications as the 555 timer recommended by Dr. Clark. The CD4069 has a very weak output relatively.

Additionally, the unit is flimsily constructed having a battery connector, a couple of alligator chips, and some electronics covered with a piece of rubber tubing. There is not any on-off switch.

If you can not afford anything better, get it, but do not expect as good of results as you would get with better zappers.

The $10 Zapper from tendollarzapper.com

This zapper is a better-made zapper for those who can not afford anything else. It does work for some problems but will not match the results of more expensive zappers.

It does use a CMOS 555-timer chip, which is better than the other cheap zappers. Also, the electronics is sealed and it comes with better cable connections. It does include an LED to show that it is on.

For the price, it is a far better $10 zapper than any other that I have examined.

ClarkZapper.com

This product is cheaply made and badly over priced. It provides only one frequency and the circuit board is loose on the inside allowing the wires to break internally. It does not offer features of units selling for $30 less.

Super Zapper Deluxe

This is an expensive but well built zapper with additional features. It has the ability to run special zapping programs that are set for specific parasites if you know which ones to use.

Of concern is the use of wrist straps, which are considered less effective than other electrodes for zapping.

Also, the additional modules for different conditions and parasites can add up to a significant cost.

Build your own zapper

Video and audio instructions are available at www.clarkzapper.net.

The Ultimate Zapper

Introduced by Ken Presner, the ultimate zapper is a low quality zapper with excessive pricing and limited flexibility. He provides a lot of hype and accuses anyone who does not like his product of being liars, etc.

It does offer a 2.5 kHz frequency, which has some advantages over 30 kHz but is probably not as good as having dual frequencies. Ken also offers footpads as an option, which can produce better results. It also offers a higher duty cycle, which is suggested to produce better results. This has not been proven in testing however. The so-called super-stabilized output causes the output to dip negative, which Dr. Clark commands not to do.

Ken also provides a wall adapter which, I have concerns about. The zapper should be operated with a 9-volt battery.

	6-Pack	CC2	UZI-3	MY-3	Termi= nator	Clark A6	Ulti mate	AutoZap	Zapper Digital	B
Kills Bacteria	Y	Y	Y	Y	?	?	?	N	?	?
Kills Protozoa	Y	Y	Y	Y	?	?	?	N	?	?
Kills Fungi	Y	Y	Y	Y	?	?	?	N	?	?
Frequencies	14	29	41	273	1	1	1	?	Note 1	1
Accuracy	1 %	0.25	0.25	0.05	10	10	10	?	?	10
Modes	6	6	8	25	1	1	1	6	1	1
Positive offset	Y	Y	Y	Y	?	Y	N	Y	?	N
Current Control	Y	Y	Y	Y	N	N	N	N	N	N
Microprocessor based	Y	Y	Y	Y	N	N	N	Y	Y	N
Rugged Construction	Y	Y	Y	Y	Y	Y	N	Y	Y	?
Output Protected	Y	Y	Y	Y	N	Y	N	Y	Y	N
Rife frequencies	Some	Some	Some	Y	N	N	N	N	???	N
Copper Paddles	Y	Y	Y	Y	N	N	N	N	N	N
Wriststraps	Opt	Opt	Opt	Opt	Y	Y	Y	Y	Y	Y
Augmentation pads	Opt	Opt	Y	Y	N	N	N	N	N	N
Battery Indicator	Y	Y	Y	Y	?	Y	N	Y	?	N
Tri-color status	Y	Y	Y	Y	N	N	N	N	N	N
Automatic cycling	N	N	N	N	N	Y	N	Y	Y	N
#Automatic cycles	0	0	Y	All	0	0	0	5	0	0
Super Stabilized Output	N	N	N	N	N	N	Y	N	N	N
Zappicator Cycle	N	N	N	N	N	N	N	Y	N	N
Constant Voltage	Y	Y	Y	Y	N	N	N	N	Y	N
Battery Life	Good	Good	Good	Good	N	OK	N	Long	?	N
Spread Spectrum	N	N	N	N	N	N	N	Y	N	N
Swept frequency	N	N	N	N	N	N	N	Y	N	N
Smart Keys	N	N	N	N	N	N	N	N	Y	N
Touch zappers	N	N	N	N	Y	N	N	N	N	N

Terminator zapper

This unit introduced by Don Croft is well built but has some limitations. This zapper provides a frequency of 15 Hz, which is very low and apparently penetrates well. The problems with this unit are that the electrodes as very small (the size of a penny) and very close together. While the 15 Hz signal does tend to compensate somewhat, it would be better if they were farther apart and larger.

Orgone Zapper

This is an improved version offered by Don Croft and also by Orgonize Africa. This zapper has the same limitations as the Terminator zapper.

Mini FG frequency generator

This is a frequency programmable zapper so that you can run sophisticated zapping programs. This zapper is also expensive, costing over $400 at the time of this writing (2005).

Gamma Full Frequency Generator

This unit is a step up from the Mini FG and is highly rated. It also costs more.

AutoZap zapper

This is a good quality unit from Arthur Doerksen that offers dual frequencies and very long battery life. You should be note that the very long battery life might result in lower signal applied to the body.

Comparing Zappers

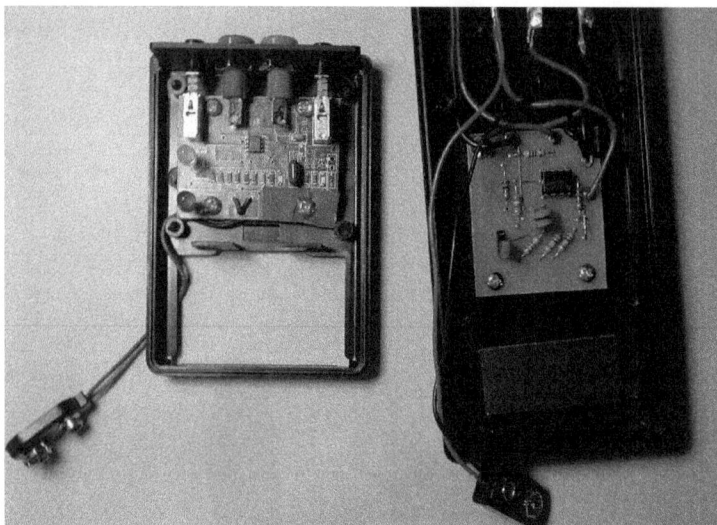

ParaZapper PLUS	Competitor's 30 kHz
Dual Frequency	Single Frequency
Tricolor status LED	Single color on LED
Secured battery cable	No restraint on cable
Low Battery indicator	No low battery LED
Quality SMT green board	Cheap board and parts
Smaller easy to store case	Bulky oversized case
Quality industrial grade components	Cheaper low precision parts
Cost $78	Cost $95-$99

Why pay more for a cheaper single frequency unit? Even out cheapest $62 unit is dual frequency.

What should your zapper have as a minimum?

Microprocessor based
Multiple modes and multiple frequencies
TLC555 timer chip as an output driver
Banana plug connectors
    (Some competitors use different connectors so only the parts they offer will work)

9 volt battery operation
Copper handle contacts ¾ diameter and 4-5 inches long

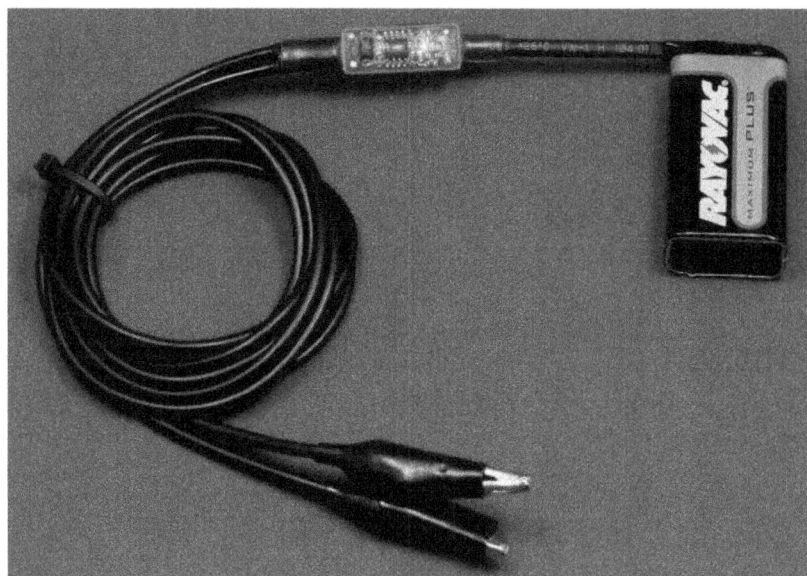

Zapper schematics
Schematic for zapperplans.com zapper

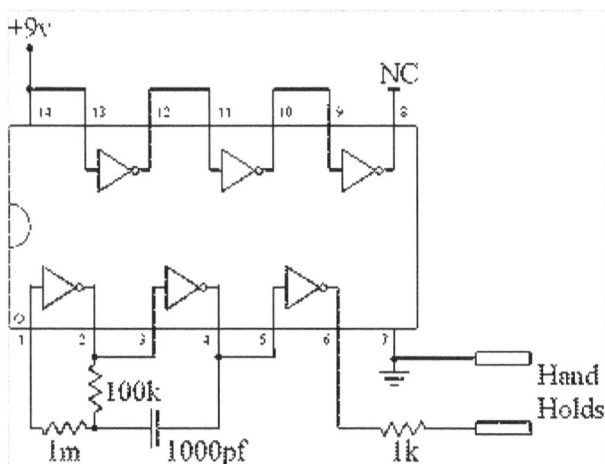

Not as good as 555 based zappers.

Dr.Hulda Clark Zapper Schematic

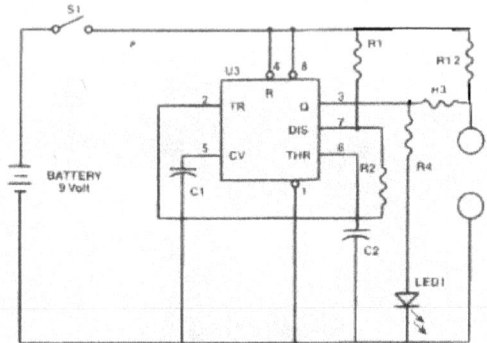

R1	1K
R2	3.9K
R3	1K
R4	4.7K
C1	.01μf
C2	.0047μf
U3	TLC555IP
LED1	2 ma LED Red
R12	39K

Latest Dr. Clark Zapper Schematic

Schematic for standard ParaZapper

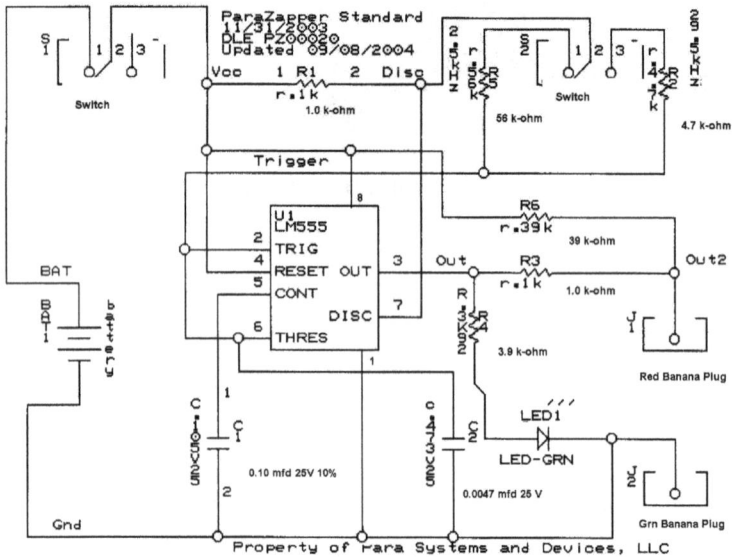

Property of Para Systems and Devices, LLC

R1	1.00k 1%	C1	0.1 mfd ceramic
R2	4.75k 1%	C2	.0047 mfd poly
R3	1.00k 1%	U1	TLC555IP
R4	3.92k 1%	J1	Banana jack RED
R5	56.2k 1%	J2	Banana jack Green
R6	39.2k 1%	SW1,SW2	SPDT Push button latch

Zapper Comparison

10 different zappers

	6-Pack	CC2	UZI-3	MY-3	Termi=nator	Clark A6	Ultimate	AutoZap	Zapper Digital	Beck
Kills Bacteria	Y	Y	Y	Y	?	?	?	N	?	?
Kills Protozoa	Y	Y	Y	Y	?	?	?	N	?	?
Kills Fungi	Y	Y	Y	Y	?	?	?	N	?	?
Frequencies	14	29	41	273	1	1	1	?	Note 1	1
Accuracy	1 %	0.25	0.25	0.05	10	10	10	?	?	10
Modes	6	6	8	25	1	1	1	6	1	1
Positiveoffset	Y	Y	Y	Y	?	Y	N	Y	?	N
CurrentControl	Y	Y	Y	Y	N	N	N	N	N	N
Microprocessor based	Y	Y	Y	Y	N	N	N	Y	Y	N
Rugged Construction	Y	Y	Y	Y	Y	Y	N	Y	Y	?
Output Protected	Y	Y	Y	Y	N	Y	N	Y	Y	N
Rifefrequencies	Some	Some	Some	Y	N	N	N	N	???	N
Copper Paddles	Y	Y	Y	Y	N	N	N	N	N	N
Wriststraps	Opt	Opt	Opt	Opt	Y	Y	Y	Y	Y	Y
Augmentationpads	Opt	Opt	Y	Y	N	N	N	N	N	N
BatteryIndicator	Y	Y	Y	Y	?	Y	N	Y	?	N
Tri-colorstatus	Y	Y	Y	Y	N	N	N	N	N	N
Automaticcycling	N	N	N	N	N	Y	N	Y	Y	N
#Automaticcycles	0	0	Y	All	0	0	0	5	0	0
SuperStabilizedOutput	N	N	N	N	N	N	Y	N	N	N
ZappicatorCycle	N	N	N	N	N	N	N	Y	N	N
Constant Voltage	Y	Y	Y	Y	N	N	N	N	Y	N
BatteryLife	Good	Good	Good	Good	N	OK	N	Long	?	N
SpreadSpectrum	N	N	N	N	N	N	N	Y	N	N
Sweptfrequency	N	N	N	N	N	N	N	Y	N	N
SmartKeys	N	N	N	N	N	N	N	N	Y	N
Touchzappers	N	N	N	N	Y	N	N	N	N	N